T0275945

SpringerBriefs in Applied Sciences and Technology

More information about this series at http://www.springer.com/series/8884

Prachi Singh · Ganesh R. Kokil
Karnaker R. Tupally · Kingshuk Poddar
Aaron Tan · Venky Venkatesan
Harendra S. Parekh · Giorgia Pastorin

Therapeutic Perspectives in Type-1 Diabetes

 Springer

Prachi Singh
School of Biological Sciences
Nanyang Technological University
Singapore
Singapore

Ganesh R. Kokil
Pharmacy Australia Centre of Excellence
The University of Queensland
Brisbane, QLD
Australia

Karnaker R. Tupally
Pharmacy Australia Centre of Excellence
The University of Queensland
Brisbane, QLD
Australia

Kingshuk Poddar
Department of Orthopedic Surgery
National University of Singapore
Singapore
Singapore

Aaron Tan
University College London
London
UK

Venky Venkatesan
Faculty of Engineering
National University of Singapore
Singapore
Singapore

Harendra S. Parekh
Pharmacy Australia Centre of Excellence
The University of Queensland
Brisbane, QLD
Australia

Giorgia Pastorin
Faculty of Engineering/Science
National University of Singapore
Singapore
Singapore

ISSN 2191-530X   ISSN 2191-5318   (electronic)
SpringerBriefs in Applied Sciences and Technology
ISBN 978-981-10-0601-2   ISBN 978-981-10-0602-9   (eBook)
DOI 10.1007/978-981-10-0602-9

Library of Congress Control Number: 2016933240

© The Author(s) 2016
This work is subject to copyright. All rights are reserved by the Publisher, whether the whole or part
of the material is concerned, specifically the rights of translation, reprinting, reuse of illustrations,
recitation, broadcasting, reproduction on microfilms or in any other physical way, and transmission
or information storage and retrieval, electronic adaptation, computer software, or by similar or dissimilar
methodology now known or hereafter developed.
The use of general descriptive names, registered names, trademarks, service marks, etc. in this
publication does not imply, even in the absence of a specific statement, that such names are exempt from
the relevant protective laws and regulations and therefore free for general use.
The publisher, the authors and the editors are safe to assume that the advice and information in this
book are believed to be true and accurate at the date of publication. Neither the publisher nor the
authors or the editors give a warranty, express or implied, with respect to the material contained herein or
for any errors or omissions that may have been made.

Printed on acid-free paper

This Springer imprint is published by Springer Nature
The registered company is Springer Science+Business Media Singapore Pte Ltd.

Contents

Overview

Type 1 diabetes (T1D), a disease that asserts through autoimmune response marked by an infiltration of pancreatic tissue with autoreactive $CD4^+/CD8^+$ T-lymphocytes, leads to comprehensive destruction of insulin producing β-cells and a loss of glucose homeostasis. The World Health Organization (2011) estimates global prevalence approaching 350 million people, and at the current pace this number is likely to double by the year 2025. Although T1D prevalence accounts for only 5–10 % of total diabetes ($\approx$35 million) cases, its increasing incidence is projected to affect a global population of 80 million by the year 2025, with the vast proportion being juveniles, below 5 years of age. Among the pathogenic factors associated with T1D are a range of autoantibodies and disease pathology further complicated by a myriad of environmental and genetic factors associated with several human and (non-) human leukocyte antigenic (HLA) genes. To date, clinicians' understanding of T1D aetiology and progression has been largely based upon the findings drawn from rodent models of the disease, however, this serves to further obscure our understanding of the disease, given that considerable disparities exist when translating the same data to human subjects. Limited availability of autoantibody assays and biomarkers and the inevitable incidence of micro- and macrovascular complications further hamper successful disease intervention. Lifelong insulin replacement remains the mainstay of symptomatic treatment and despite significant advances being made over recent decades with interventions such as closed-loop glucose monitoring, β-cell replacement, immunotherapy with antiglobulins and regulatory T-cells researchers are yet to find a safe, effective and universally acceptable approach for curing T1D. In this Review we offer critical insights and appraisal of recent breakthroughs in T1D modulation, with particular emphasis on the potential impact of current prevention and treatment strategies, closing with discussion on recent successes and failures in clinical trials.

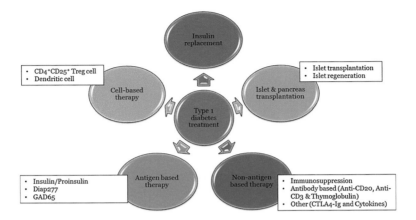

Chapter 1
Introduction to Diabetes and Type 1 Diabetes

1.1 Introduction

1.1.1 Defining Diabetes

The World Health Organization (WHO) classification of diabetes has been the subject of subtle iterations over the years, with their most recent classification published back in 1985. At that time an expert committee on the diagnosis and classification of diabetes mellitus identified two main types of diabetes: type 1 diabetes (T1D) and type 2 diabetes (T2D). Additionally, the less common gestational diabetes (GDM) that occurs exclusively during pregnancy was also observed and noted. Higher physiological levels of glucose, often called hyperglycemia, resulting from either insulin resistance or deficiency are a key characteristic of T2D (Alberti and Zimmet 1998; Cnop et al. 2005). In the case of GDM expectant mothers, those usually in their third trimester are more susceptible to persistently high blood glucose levels and this typically affects 3–10 % of the pregnant population (Buchanan and Xiang 2005; Centers for Disease Control and Prevention, 2005; Setji et al. 2005; Singh and Rastogi 2008; Ornoy 2011). For the purpose of this book, however, we project our focus primarily on the wide range of issues and challenges common to T1D. Table 1.1 provides a snapshot of the key features of each diabetes classification alluded to above, namely the prevalence, etiology, symptoms, diagnostic indicators, and associated complications.

Insulin, an endogenous hormone is secreted by b-cells present in the Islets of Langerhans of the pancreas and is responsible for maintaining homeostatic, euglycemic levels (80–110 mg/dL or 4.5–6.2 mmol/L) within the body (Fig. 1.1). Lowered insulin secretion elevates blood glucose levels beyond the normal range, leading to hyperglycemia. Insulin facilitates glucose utilization, energy generation and stores energy, predominantly in the hepatocytes, adipocytes, and myocytes. The

© The Author(s) 2016
P. Singh et al., *Therapeutic Perspectives in Type-1 Diabetes*,
SpringerBriefs in Applied Sciences and Technology,
DOI 10.1007/978-981-10-0602-9_1

Table 1.1 Classification of diabetes and their associated attributes

Classification of diabetes	Type 1 diabetes (T1D)	Type 2 diabetes (T2D)	Gestational diabetes
Synonym	IDDM, juvenile-onset diabetes	NIDDM, adult-onset diabetes	GDM
Epidemiology	5–10 % of total cases	90–95 % of total cases	3–10 % of pregnancies
Risk group	Typically <20-years old	Adults (≥25-years)	Pregnant women >25-years old
Risk factors	Susceptible HLA genes, autoantibodies, environmental factors (dietary factors and viral infection)	Obesity, old age, and unhealthy lifestyle	Placental hormones, elevated BMI before pregnancy, age of conception, and race
Symptoms	Polydypsia, polyphagia, polyuria and weight loss	Polydypsia, polyphagia, polyuria and weight loss, fatigue, slow wound healing	As for T1D and T2D
Etiology	Autoimmune β-cell destruction, idiopathic	Insulin resistance	Insulin resistance or deficiency
Diagnosis	HbA1c level >6.5 %, GAD, insulin and IA2 antibodies and HLA genotyping	Fasting BGL ≥7 mmol/L, 2 h BGL ≥7.8 mmol/L	OGTT (75 g glucose load, fasting BGL ≥7 mmol/L, 2 h BGL ≥7.8 mmol/L)
Complications	Micro- and macrovascular diseases. Prominence of microvascular (e.g., retinopathy, nephropathy, and neuropathy)	Micro- and macrovascular diseases. Prominence of macrovascular (e.g., heart disease, hyperlipidemia and ketoacidosis)	Congenital defects, neonatal hypo/hyperglycemia, fetal macrosomia, and perinatal mortality

Body mass index = BMI; Insulin-dependent diabetes mellitus = IDDM; Non-insulin dependent diabetes mellitus = NIDDM; Glutamic acid decarboxylase = GAD; HbA1c = Glycated hemoglobin; Oral glucose tolerance test = OGTT; Islet antigen = IA2

hormone is first synthesized in its precursor form which is known as 'preproinslin', after which proteolytic enzymes cleave the signal sequence and generate 'proinsulin' (86 amino acid containing protein) arranged in three distinct peptide chains: A- (21 amino acid), C- (31 amino acid), and B-chain (30 amino acid). The protein is further processed by proteases within endoplasmic reticulum (ER), cleaving the C-peptide (C-chain) and forming a 51 amino acid bridged peptide, which is active insulin. The discarded C-peptide portion of proinsulin is a reliable marker of insulin secretion and is routinely used as a diagnostic marker, as well as to develop therapies and better understand the pathophysiology of diabetes mellitus.

T1D is an autoimmune disease, usually diagnosed in children and young adults, where pancreatic β-cells are attacked by autoantigens, that is, an immune-mediated

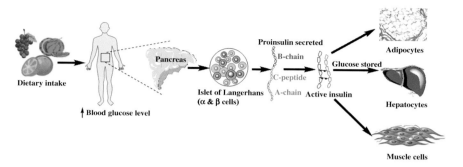

Fig. 1.1 General scheme of insulin biosynthesis and physiological function

attack against one's own tissues, specifically, insulin producing β-cells in the Islets of Langerhans of the pancreas resulting in depleted insulin levels, which, if left untreated, leads to hyperglycemia. There is substantial credible evidence highlighting that autoimmunity is the root cause of T1D, this stemming from the gross infiltration of Islet of Langerhans by antigen presenting cells (APCs) such as macrophages (Mφ), dendritic cells (DCs) as well as B- and T-lymphocytes (CD4$^+$ and CD8$^+$) (Nerup et al. 1994; Kaufman 2003; Pihoker et al. 2005; Kabelitz et al. 2008).

The human adaptive immune system possesses a characteristic feature to respond to the foreign/non-self-antigens, wherein it fails to react to antigen derived from self-tissues (self-antigens/autoantigens). However, for unknown reasons, the immune system mistakenly identifies these autoantigens as foreign bodies and mounts an immune response against them. Once these autoantigens are identified, peptide major histocompatibility complex (pMHC) expressed on the surface of APCs (DCs and mTECs; medullary thymic epithelial cells) within thymic medulla complexes with the T-cell receptor (TCR) on the T-cell's surface (shown in Fig. 2.2) is given but artwork not provided. Please check and provide artwork and caption for Fig. 3 or delete these citation." –>) (Concannon et al. 2009). Even after the identification of these autoantigens, a vast population fails to develop autoimmunity due to active self-tolerance mechanisms (Parish and Heath 2008). In a state of active tolerance, autoreactive T-cells developed toward autoantigens are then purged out of the system either by (a) *central/thymic tolerance* or (b) *peripheral tolerance*. Central tolerance occurs within the thymus gland wherein T-cells undergo positive and negative selection. In positive selection, T-cells are given survival signal following their binding with MHC/peptide complex with adequate affinity. During development within thymus, T-cells are educated to delete T-cells with high avidity T-cell receptor (TCR) for autoantigens (self-reactive). These self-reactive T-cells are purged either by; (1) *clonal deletion*, i.e., physiological deletion of the autoreactive T-cells, or (2) *anergy*, i.e., functional inactivation of the autoreactive T-cells. On the other hand a minor population of low avidity autoreactive T-cells often manages to survive and escape central tolerance processing, migrating into the periphery. These low avidity T-cells in the periphery are of particular concern given their ability to be processed by the host immune system are a danger to the host immune system, with the potential of developing autoimmunity. Although these autoreactive T-cells bypass the central tolerance, they can still be curbed from spreading within the periphery by what is known as *peripheral tolerance* (Mueller 2009). One of the purported regulatory actions of peripheral tolerance is *self-ignorance,* whereby autoreactive T-cells persist in the periphery but in a naïve state, where they are incapable of imparting any deleterious autoimmune effects. However, it has been shown in a range of experimental models that, if 'self' pMHC complexes are sufficiently outnumbered, the low avidity T-cells may be activated, once again leading to the genesis of autoimmunity (Zehn and Bevan 2006; Henrickson et al. 2008).

1.1.2 Type 1 Diabetes: Epidemiology and Perspectives of a Global Epidemic

According to recent World Health Organization (WHO) estimates, nearly 350 million patients globally suffer from either form of diabetes, with as many as 35 million cases, or 10 % of the total, being attributed to T1D. Furthermore, if WHO predictions are to be believed by then this number is expected to be more than double by the year 2025 ($\approx$80 million cases) (Haller et al. 2005; Centers for Disease Control and Prevention 2011). Population-based registries maintained over a span of the last 50 years have shown a global variation in the incidence, prevalence, and temporal trends of T1D. A highly cited article by Onkamo et al. provided one of the most informative and invaluable global incidence trends of T1D; this represents the most extensive analysis, collating data from 37 independent studies on the alarming global increase (across 27 countries) in T1D cases, over almost four decades (1960–1996) (Onkamo et al. 1999). The accumulated data indicated an annual increase in incidence of 3 % with a greater relative increase in lower-incidence countries. This in retrospect proved relatively conservative with the actual increase now being estimated close to 3–5 % (Green and Patterson 2001; Patterson and Dahlquist 2009).

Global prevalence and trends of T1D were, up until 1980 particularly difficult to map given the dearth of standardized epidemiological data collection and analysis methods. However, this situation changed with the commencement of large-scale epidemiological studies such as DIAMOND and EURODIAB (Karvonen et al. 2000; Green and Patterson 2001). Diabetes Mondiale (abbreviated to DIAMOND), a comprehensive multinational project was initiated in 1990 by the WHO; their study incorporated 114 independent data sets of children only (<14 years of age) collated from 112 centers across 57 countries from 1990 to 1999. An annual increase in incidence of 2.8 % was calculated over the study period, which was in-line with previous estimates (Karvonen 2006). A second significant study, EURODIAB project (epidemiology and prevention of diabetes) was conducted primarily across the regions of Europe and Israel from 1988 to 2003 with the inclusion of nearly 17 million children (<15 years of age). An overall annual increase of 3.9 % calculated from the EURODIAB study was somewhat higher than the projections from DIAMOND and Onkamo's studies. A closer look at the EURODIAB data revealed the highest increase in incidence rates for the youngest age group of 0–4 years (5.4 %), compared with 4.3 and 2.9 % in the 5–9 and 10–14 age groups, respectively. The studies collectively reported no significant difference between male and female incidence risk, with peak incidence occurring at puberty. Following analysis of the data from the EURODIAB study, disease prevalence was projected to increase by 70 % over 15 years, from 94,000 cases in 2005 to 160,000 cases by 2020 (Patterson and Dahlquist 2009). Indeed such a steep increase cannot be explained by genetic predisposition alone, as this could not be mapped over a much longer time frame.

The aforementioned studies have not only proven invaluable to predict the overall incidence trend, but additionally they highlight stark variations in prevalence across the continents of Europe and North America compared to Asia. Notwithstanding the significant trend differentials it remains uncertain whether this difference is absolute or attributed to the sparse reporting from within the Asian regions. Furthermore, there are reports of intra-country variations coupled with north–south hemispheric divide, in T1D incidence. Sardinia for example, had a five-fold greater incidence compared to the other mainland Italian regions (Soltesz et al. 2007; Maahs et al. 2010), while countries such as China and Venezuela presented incidence rates as low as 0.1 cases/100,000/year compared to a staggering 40.9 cases/100,000/year in Finland, which correlates to a staggering *circa* 400-fold difference (Karvonen et al. 2000; Karvonen 2006; Harjutsalo et al. 2008). Another prospective USA-based study launched in 2000, which concludes in 2015, is the Search for Diabetes in Youth (SEARCH) project. This study, encompassing subjects <20-years old is unique as it is the first such targeted epidemiological project including patients of all major races/ethnic groups in the USA and will study the incidence of T1D as well as attempting to identify key associated environmental factors of the disease.

The findings of studies surmised above all paint an alarming picture on T1D global prevalence. Further, it is predicted that future projected incidence of T1D will be higher in presently reported low incidence countries, as well as in developing countries, due to their adoption of western nutrition and lifestyle habits. Collectively, the data depict that steep increase in the incidence is attributed to a broad range of factors aside from genetic predisposition, i.e., environmental factors that include viral infection, increased maternal age, early life nutrition habits, etc. All these factors have driven epidemiologists and the wider research community to carefully scrutinize possible causes for the sharp increases in T1D cases, along with the observed variations in different populations, all of which is expected to place serious pressure on already underfunded healthcare budgets and operations, and this can only be alleviated through better regulated, globally coordinated epidemiological studies in T1D.

References

Alberti KG, Zimmet PZ (1998) Definition, diagnosis and classification of diabetes mellitus and its complications. Part 1: diagnosis and classification of diabetes mellitus provisional report of a WHO consultation. Diabet Med 15(7):539–553

Buchanan TA, Xiang AH (2005) Gestational diabetes mellitus. J Clin Invest 115(3):485–491

Centers for Disease Control and Prevention (2005) National diabetes fact sheet: general information and national estimates on diabetes in the United States. U.S. Department of Health and Human Services, Atlanta, GA

Centers for Disease Control and Prevention (2011) National diabetes fact sheet: general information and national estimates on diabetes in the United States. U.S. Department of Health and Human Services, Atlanta, GA

Cnop M, Welsch N et al (2005) Mechanisms of pancreatic beta-cell death in type 1 and type 2 diabetes: many differences, few similarities. Diabetes 54(Suppl 2):S97–107

Concannon P, Rich SS et al (2009) Genetics of type 1A diabetes. N Engl J Med 360(16): 1646–1654

Green A, Patterson CC (2001) Trends in the incidence of childhood-onset diabetes in Europe 1989–1998. Diabetologia 44(Suppl 3):B3–8

Haller MJ, Atkinson MA et al (2005) Type 1 diabetes mellitus: etiology, presentation, and management. Pediatr Clin North Am 52(6):1553–1578

Harjutsalo V, Sjöberg L et al (2008) Time trends in the incidence of type 1 diabetes in Finnish children: a cohort study. Lancet 371(9626):1777–1782

Henrickson SE, Mempel TR et al (2008) T cell sensing of antigen dose governs interactive behavior with dendritic cells and sets a threshold for T cell activation. Nat Immunol 9(3):282–291

Kabelitz D, Geissler EK et al (2008) Toward cell-based therapy of type 1 diabetes. Trends Immunol 29(2):68–74

Karvonen M (2006) Incidence and trends of childhood Type 1 diabetes worldwide 1990–1999. Diabet Med 23(8):857–866

Karvonen M, Viik-Kajander M et al (2000) Incidence of childhood type 1 diabetes worldwide. Diabetes Mondiale (DiaMond) Project Group. Diabetes Care 23(10):1516–1526

Kaufman FR (2003) Type 1 diabetes mellitus. Pediatr Rev 24(9):291–300

Maahs DM, West NA et al (2010) Epidemiology of type 1 diabetes. Endocrinol Metab Clin North Am 39(3):481–497

Mueller DL (2009) Mechanisms maintaining peripheral tolerance. Nat Immunol 11(1):21–27

Nerup J, Poulsen TM et al (1994) On the pathogenesis of IDDM. Diabetologia 37(2):S82–S89

Onkamo P, Väänänen S et al (1999) Worldwide increase in incidence of Type 1 diabetes—the analysis of the data on published incidence trends. Diabetologia 42(12):1395–1403

Ornoy A (2011) Prenatal origin of obesity and their complications: Gestational diabetes, maternal overweight and the paradoxical effects of fetal growth restriction and macrosomia. Reprod Toxicol 32(2):205–212

Parish IA, Heath WR (2008) Too dangerous to ignore: self-tolerance and the control of ignorant autoreactive T cells. Immunol Cell Biol 86(2):146–152

Patterson CC, Dahlquist GG (2009) Incidence trends for childhood type 1 diabetes in Europe during 1989–2003 and predicted new cases 2005–20: a multicentre prospective registration study. Lancet 373(9680):2027–2033

Pihoker C, Gilliam LK et al (2005) Autoantibodies in diabetes. Diabetes 54(Suppl 2):S52–S61

Setji TL, Brown AJ, Feinglos MN (2005) Gestational diabetes mellitus. Clin Diabetes 23(1):17–24

Singh S, Rastogi A (2008) Gestational diabetes mellitus. Diab Metab Syndr: Clin Res Rev 2(3):227–234

Soltesz G, Patterson CC et al (2007) Worldwide childhood type 1 diabetes incidence—what can we learn from epidemiology? Pediatr Diabetes 8(Suppl 6):6–14

Zehn D, Bevan MJ (2006) T cells with low avidity for a tissue-restricted antigen routinely evade central and peripheral tolerance and cause autoimmunity. Immunity 25(2):261–270

Chapter 2
Triggers Causing Type 1 Diabetes

2.1 Introduction

The precise cause(s) of T1D are not fully understood to the researchers; however, it appears to be a combination of several factors which includes genetics as well as environment. It is observed that children with parents or sibling with T1D has 2–6 % risk compared to a risk of about 0.4 % in general population. Other autoimmune diseases, such as thyroid disease and celiac disease make a patient more prone to develop T1D. Furthermore, it is observed that certain ethnicities across the world are at greater risk to develop the disease compared to other. For example: Caucasians in America are more susceptible compared to African Americans, Native Americans, Asian Americans, and Latinos (Nielsen et al. 2014).

2.1.1 Genetic Susceptibilty

The significant role of ones' genetic makeup in T1D has long been established and it is for this reason that a great deal of investigation has focused on identifying susceptible gene families. The best evidence for the involvement of genetic components comes from the studies conducted and correlating the data from both animal models of the disease and humans (Risch 1987; Davies et al. 1994; Lyons and Wicker 1999).

© The Author(s) 2016
P. Singh et al., *Therapeutic Perspectives in Type-1 Diabetes*,
SpringerBriefs in Applied Sciences and Technology,
DOI 10.1007/978-981-10-0602-9_2

2.1.1.1 Human Leucocyte Antigen (HLA Genes)

Most T1D risk is contributed by major histocompatibility complexes (MHC), located on the short arm of chromosome 6 (locant 6p21.3), otherwise termed human leucocyte genes (HLA) (Pociot and McDermott 2002). HLA genes fall into three different families, class I, II, and III (Fig. 2.1). Among the three families, HLA class II genes, with specific allelic regions *DR* and *DQ* are primarily involved in T1D progression (Table 2.1). These genes together constitute around 40–50 % of predisposition risk. Within families the risk of disease is found to be highest in monozygotic twins, followed by first and second-degree relatives with 50 and 25 %

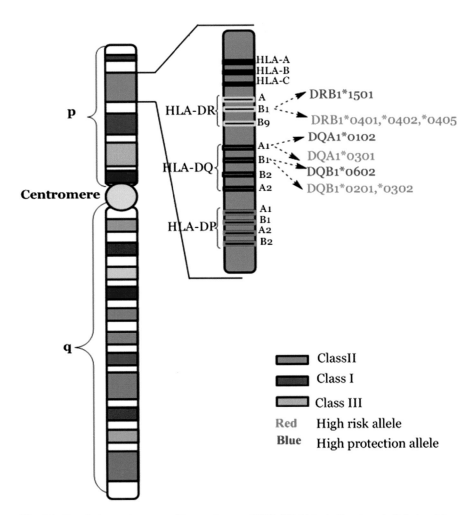

Fig. 2.1 Detailed genetic composition on human MHC (HLA) including type 1 diabetes risk

Table 2.1 Role of non-HLA genes in susceptibility

Non-HLA genes and chromosome location	PTPN22 (1p13.3–13.1)	INS (11p15.5)	CTLA-4 (2q33.2)	IL2RA/CD25 (10p15.1)
Role in T1D	Encodes LYP; a dominant negative antigen-TCR regulator that acts via dephosphorylation and inactivation of T-cell	Encodes pre-proinsulin peptide (insulin precursor)	Surface molecule present on T-cells; negative regulator of TCR activation; blocks B7-1 and B7-2 (also called CD80 & CD86) (see Fig. 2.2)	Encodes α-chain of IL-2 receptor complex (also called CD25) essential for self-tolerance of T_{reg} cells
Major findings/Mutations	SNP at base pair 1858 from cysteine to thymine (1858C → 1858T) resulted in amino acid change from arginine to tryptophan (R620 W) Mutation found higher in T1D population compared to healthy controls	VNTR regulates INS expression; VNTR classification: • Class I: 2–5 fold increase in T1D risk (26–63 repeats/~0.57kbp) • Class II: Most rare allele (80–85 repeats/~1.64kbp) • Class III: Increased expression in thymus and decreased in pancreas and results in efficient thymus selection and protects against T1D (141–209 repeats/~2.4kbp)	Threonine to alanine (A49G) substitution lead to abnormal post-translational glycosylation and decreased CTLA-4 surface expression and reduced T_{reg} cells. Cytosine to thymidine (C318T) leads to increased CTLA-4 cell surface expression and hence imparts protection against T1D	Different IL-2R SNP have been associated with high risk of T1D such as rs41295061 and ss52580101; rs12722495 found to alter T_{reg} function
References	Bottini et al. (2004, 2006), Nielsen et al. (2011)	Bell et al. (1984), Kennedy et al. (1995), Pugliese et al. (1997), Vafiadis et al. (1997)	Salomon and Bluestone (2001), Anjos et al. (2002), Wang et al. (2002), Buhlmann et al. (2003), Wang et al. (2011)	Espino-Paisan et al. (2011), Gillespie and Boraska (2011), Garg et al. (2012)

Protein tyrosine phosphatase non-receptor 22 = PTPN22; Insulin gene = INS; Cytotoxic T-lymphocyte antigen 4 = CTLA-4; Interleukin 2 receptor alpha = IL2RA; Lymphoid tyrosine phosphatase = LYP; T-cell receptor = TCR; Single nucleotide polymorphism = SNP; Variable number of tandem repeats = VNTR; kilo base pair = kbp; T regulatory cells = T_{reg}

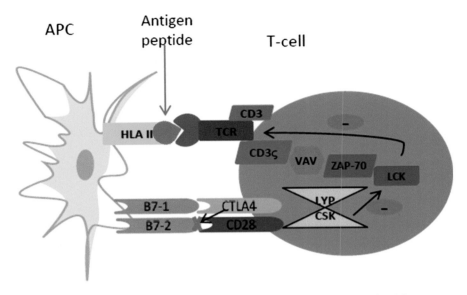

Fig. 2.2 Communication and interaction pathways between an APC and T-cell. APC presents a processed antigen peptide via HLA II on to the TCR/CD3 complex present on the T-cell surface. For the full activation of the T-cell, a second signal is required and this signal is generated via costimulatory molecular interaction between CD28 on the T-cell surface and B7-1 (CD80) and B7-2 (CD86) on the APC surface. Additional signaling messengers are present in the cell cytosol with positive regulatory tyrosine kinases such as LCK, CSK, VAV, ZAP-70 and CD3-ζ. These tyrosine kinases are inhibited by the LYP/CSK complex. CTLA-4 molecule is an inhibitory molecule, which competes with CD28 for occupying the B7-1 or B7-2 pockets; this inhibits the required secondary or costimulatory signal for T-cell activation leading to quiescent or apoptotic T-cells (Illustration modified from Wiebolt et al. 2010)

sharing, respectively. In contrast, several HLA class II alleles have been shown to confer a higher degree of protection against the development of T1D. One such gene is encoded at *HLA-DQB1*0602* (Todd et al. 1987). It is further interesting to note that these alleles are capable of protection even in the presence of high risk alleles or T1D associated autoantibodies (Pugliese et al. 1995; Pociot et al. 2010).

2.1.1.2   Non-Human Leucocyte Antigen (Non-HLA Genes)

HLA genes account for some 40–50 % of genetic predisposition, suggesting that supplementary risk factors also play a near equivalent role in T1D progression. Efforts to identify such non-HLA genes have been possible, most notably through Genome-wide association studies (GWAS) and Gene-linkage studies (Davies et al. 1994; Hirschhorn and Daly 2005). GWAS is an approach which involves scanning markers, which can rapidly scan across the genome of an individual to find genetic variations which are associated with a particular disease (National Human Genome

Research Institute 2014). Genetic linkage studies are used to identify regions of the genome that contain genes that predispose an individual to disease. Linkage analysis is used to map genetic loci by use of observations of related individuals (Dawn and Barrett 2005). Analysis of the available data suggests that non-HLA genes have relatively smaller yet significant effect on individual risk. A list of non-HLA risk associated genes with their chromosomal location, role in T1D and mutations are highlighted in Table 2.1.

Targeting the protective genes is considered a valuable approach to T1D therapy. However, there are various barriers before their successful translational potential can be harnessed. For example, there is a need for meticulous mechanistic studies to differentiate the dual role of several genes such as PTPN22, which is purported to protect against tuberculosis (TB), while predisposing patients to both T1D and rheumatoid arthritis (RA). Additional barriers include inadequate number of study sample sets and mixed ethnic groups that further add to the complexities and our understanding of dominant functionalities in T1D.

2.1.2 Environmental Triggers

The unexpected peak in the number of T1D patients across the globe represent strong evidence that the increase cannot be due to genetic susceptibility alone, as genetic changes present over a significantly longer period of time. This suggests involvement of factors, distinct from genetics such as environmental triggers or precipitating agents in the etiology of T1D (Åkerblom and Knip 1998). The hypothesis is supported by confounding geographic variation observed within the European population with a mere 3.2 cases/100,000/year reported in Macedonia, as opposed to a staggering 40.9 cases/100,000/year in Finland (Karvonen 2006). Moreover, migratory studies have shown higher incidence of T1D in individuals, who relocated from regions with low incidence to those with higher incidence (Tull et al. 1992). With this in mind, we next discuss the epidemiological data from various studies, which convince us of the association of several putative environmental triggers such as viruses, cow's milk, vitamin D, and toxins (N-nitroso compounds) with T1D (Knip et al. 2005).

2.1.2.1 Viruses

The very first idea that manifestation of viral infections is linked to T1D etiology dates back to 1920s, where T1D onset was observed to follow a seasonal pattern (Adams 1926). Since then various viral candidates have been investigated for their role in the precipitation of the disease, such as enteroviruses (EVs) most notably coxsackie B virus (CBV), rubella, mumps, rotavirus, cytomegalovirus, and Epstein Barr virus (EBV). However, the evidences of association of viruses with T1D hitherto, have either been weak, irreproducible, and sometimes even been rejected by the scientific world due to the contradictions in the reports (Coppieters et al. 2012). Conversely, there are also a range of viruses which demonstrate a protective effect against autoimmune development, although they are deemed outside the remit of this book.

Enteroviruses (EVs) are among the most robustly reported infectious agents having a close association with T1D. EVs belong to the picornavirus family, these being small, naked, icosahedral, single stranded RNA-containing viruses. The first EV-T1D correlation was reported in 1969 (Gamble et al. 1969) while Oikarinen et al. only recently succeeded in isolating EVs from pancreatic biopsies of T1D patients (Oikarinen et al. 2008). Approximately 75 % (n = 12) of T1D patients versus 10 % (n = 10) in control subjects were found to have EV in their gut mucosa. Apart from only a handful of investigations reporting infants with increased HLA susceptibility toward T1D, several other studies present a contradictory view of the EV-T1D link. For instance, a prospective study conducted in Finland (DIPP) versus Colorado, USA (Diabetes Autoimmunity Study in the Young, DAISY) and Germany delivered contrasting verdicts with no report of increased T1D risk due to EV infection (Füchtenbusch et al. 2001; Graves et al. 2003; Salminen et al. 2003). However, it is difficult to confirm, with any level of precision whether the reported events are a result of differences in the methodology employed, or if in fact there is an actual difference, within the screened populations. Further, these studies lead to the corollary that there is a need to first identify whether the viral infections are acute or persistent in nature (Tauriainen et al. 2011) as well as assessing if the relationship found in some cases is really a causal effect or just a secondary event in patients already presenting clinical diabetes. It is interesting to note that parallel studies on another strain of coxsackie viruses, i.e. CBV3, resulted in delayed onset of diabetes due to expansion of $CD4^+CD25^+$ T_{reg} (regulatory T-cells) cells (Viskari et al. 2000, 2005). In other words, time of infection and inoculation dose have a significant role to play in the development of T1D.

Two different mechanisms have been proposed to explain the causative role of viruses in T1D. First, "*molecular mimicry*" which is described by resemblance of the P2-C protein sequence of the coxsackie virus with that of glutamic acid decarboxylase 65 (GAD65) epitope, a major antigen determinant of T1D (Hou et al. 1994). It was observed that T-cells from patients with higher risk of T1D, which responded to GAD65 determinants, also responded to the P2-C peptide of the virus (Helgason and Jonasson 1981; Schloot et al. 2001). However, autoreactive human T-cell clones that are specific for GAD65 epitope failed to proliferate poststimu-lation with the viral epitope (Schloot et al. 2001). Further studies in rat insulin-promoter lymphocytic choriomeningitis virus (RIP-LCMV) mice suggested that in order to initiate viral infection, it is necessary to have a complete homology between viral and β-cell antigens, since a mere single amino acid change flanking a cytotoxic T-lymphocyte (CTL) epitope was found to interfere with the development of T1D (Fujinami et al. 2006; Knip et al. 2010). The second hypothesized mech-anism involved in the viral etiology of T1D is "*bystander activation*". It is proposed that this bystander mechanism works via a TCR-independent mechanism (Jun and Yoon 2003). The substitute mechanisms contain activation through soluble factors or membrane-bound molecules that bind to receptors other than TCR. Several other mechanisms include the alteration of the host immune system followed by viral infection.

These proposed mechanisms are far from being able to explain all the observations. To further confuse matters there have been additional findings suggesting an inverse relationship between EVs and T1D incidence. In Finnish and Swedish studies, a decrease in the frequency of EV infection has been reflected by increased incidence of T1D compared to other countries, such as Estonia and Russia; this was later explained by the "hygiene hypothesis" (Viskari et al. 2005). Conversely, it has also been observed that childhood immunization against certain diseases of viral origin provide a protective effect against T1D (Classen and Classen 1999, 2002).

2.1.2.2 Dietary Factors

Viral infection and genetic susceptibility are not the only culprits triggering T1D. Rather, recent changes in early life diet and feeding habits implicate several dietary components as additional triggers driving autoimmune T1D. The following section discusses some of the known dietary factors linked to T1D etiology.

2.1.2.3 Early Dietary Habits—Cow's Milk and Gluten

The early exposure to cow's milk and the nutritional benefits of breast milk have been extensively investigated with outcomes having been reported. Indeed, a study amongst the Scandinavian population, showed an inverse relationship between breastfeeding and T1D incidence (Borch-Johnsen et al. 1984), while a Finnish study reported that short-term breastfeeding followed by early introduction of a cow's milk-based infant formula predisposed young children to progressive signs of β-cell autoimmunity, rendering these children genetically susceptible to T1D (Kimpimäki et al. 2001).

Other investigation focused on the role of cow's milk with supplementation from various dietary components, such as early introduction of solids and gluten containing diets. Gluten is a protein principally found in wheat and it can be further divided into two subgroups of protein, the gliadins and glutenins. Anecdotal reports claim a low incidence of T1D in rodents with diets containing low levels of protein such that inclusion of gliadin by as little as 1 % by weight, of individuals total food consumption showed T1D incidence of about 35 % compared to the 15 % incidence in control group fed on a semi-synthetic diet (Elliott and Martin 1984). Separately, two prospective cohort-based studies arrived at a conclusion that a "critical period of exposure" to cereals is associated with the progression of T1D. Norris et al. found that children exposed to cereals from 0 to 4 months of age and those exposed at an age of 7 months or older have higher risk of developing autoimmunity with hazard ratio (i.e., hazard in the exposed group/hazard in the unexposed group) (HR) of 4.32 and 5.36, respectively (Norris et al. 2005). A German study reported similar results with a five-fold (HR of 4.0) higher risk of T1D development in children exposed to a gluten containing diet before age of 3 months compared to those exposed after the age of 6 months (Ziegler et al. 2003). The risk associated with the early exposure may be due to anomalous immune response to cereal antigens in an immature gut system of susceptible individuals, while the late exposure risk in older children could be attributed to the consumption of larger portions of cereals.

The inferences from the above studies are limited by numerous methodological incongruences, such that discrepancies in the reported outcomes may well be attributed to differences in the study population or to the variation in the feeding practices across the countries where the studies were conducted. Hence it is imperative to address this issue systematically, as attempted in the 'Trial to Reduce Insulin-dependent diabetes mellitus (IDDM) in the Genetically at Risk (TRIGR).' In a recent, albeit limited scope pilot study, and the first of this kind in humans, a link suggesting that it is possible to manipulate spontaneous β-cell autoimmunity by nutritional intervention during infancy was reported (Åkerblom et al. 2005). TRIGR, an international group conducting collaborative clinical trials at 77 clinical centers in 15 different countries across 3 continents, began addressing the hypothesis of whether weaning with a hydrolysate protein-containing diet decreased T1D incidence in a genetically susceptible population, compared to that of a cow's milk protein diet (Åkerblom 2011; Knip et al. 2014). Successful collaborative data from the trial is expected to be major stepping stone toward primary prevention and intervention strategies of T1D.

2.1.2.4 Nitroso Compounds

There is circumstantial evidence in humans which shows a positive trend between T1D and nitrate intake (Helgason and Jonasson 1981). The most common sources of nitrates and nitrites are in drinking water and some foods, most notably smoked and cured meats and vegetables. Dietary nitrites and nitrates readily combine with endogenous amine and amides to form nitrosamines and nitrosamides. These nitroso compounds bear a structural resemblance to streptozotocin, a compound used to induce diabetes in rodent models of the disease. A study by Helgason et al. in the 1980s, showed a plausible link between consumption of smoked mutton containing N-nitroso compounds and an increased incidence of T1D; this was accompanied by a striking observation that incidence was higher in the progeny rather than the consumer (Helgason and Jonasson 1981; Helgason et al. 1982). Similarly, in a separate Finnish case-control study, strong correlation was found between intake of nitrite and nitrates obtained from drinking water and food in mothers and in their young offspring (Virtanen et al. 1994).

2.1.2.5 Vitamin D

In contrast to the above mentioned dietary factors, responsible for augmenting T1D progression, there exists considerable epidemiological, experimental, and cross-sectional evidence supporting the protective effect of vitamin D in T1D (Kriegel et al. 2010). However, human epidemiological data and interventional data supporting this school of thought are somewhat tenuous to support the conclusion. The pinnacle in T1D incidence has been linked to geographical location and may well be dictated by variations in exposure to sunlight, which differs markedly depending on geographical latitude. Furthermore, a seasonal pattern where an increased number of T1D cases arise during winter suggests an inverse correlation between exposure to sunlight and T1D. Vitamin D is predominantly supplied through ones diet although a high amount is synthesized within the skin on exposure to UVB light (detailed synthesis scheme is depicted in Fig. 2.3). Vitamin D is essential for normal insulin secretion (Zeitz et al. 2003), exerting various effects on the immune system, leading to tolerance generation and anergy, as opposed to immune activation (detailed in Fig. 2.4) (Mathieu and Badenhoop 2005). The role of vitamin D in pancreatic β-cell function may be facilitated by the binding of circulating vitamin D to the vitamin D receptor (VDR) (Hewison et al. 2003).

Various studies in non-obese diabetic (NOD) mice and biobreeding (BB) rats have shown that deficiency of vitamin D in early life results in impaired glucose tolerance and a multicenter trial in different regions of Europe has consistently shown the beneficial effect of vitamin D supplementation during the first year of life. As an expansion to the earlier mentioned EURODIAB project, another study investigated whether the use of dietary cod liver oil or other vitamin D supplement, either by the mother during pregnancy or by the infant during the first year of life, were associated with a lower risk of T1D among children (Group 1999). A strong negative association was found between maternal intake of vitamin D or cod liver oil during pregnancy and T1D diagnosis (Stene et al. 2001). Similar results were obtained from a subsequent study, where supplementation with vitamin D during pregnancy translated to relatively low risk of developing autoimmune T1D (Brekke and Ludvigsson 2007). Additionally, the beneficial effects were limited in their duration with autoimmunity declining in 1 year age group but not in the 2.5 year one. The apparent reduction in disease progression may be attributed to the immunomodulatory effects of vitamin D (Zipitis and Akobeng 2008). Here, the chronic administration of pharmacological doses of vitamin D as a primary pre-vention strategy was found to decrease insulitis and diabetes in NOD mice (Mathieu et al. 1992), although hypercalcaemia was a major side effect observed during the study as a result from enhanced reabsorption of calcium from the intestine and elevations in bone resorption (Mathieu and Badenhoop 2005; Peechakara and Pittas 2008; Danescu et al. 2009). In addition, apart from the environmental role of vitamin D, various allelic variations of VDR found on chromosome 12 are also documented associated risks of T1D (Cooper et al. 2011).

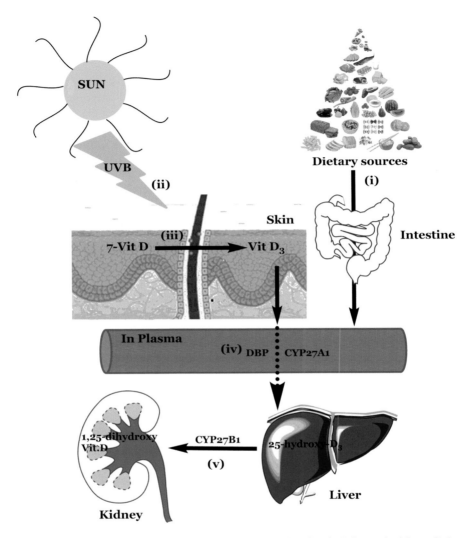

Fig. 2.3 Graphical representation of vitamin D biosynthesis. Vitamin D is acquired from (**i**) the diet or (**ii**) synthesized within skin on exposure to UVB (297-309 nm) causing photolysis of (**iii**) 7-Vitamin D (7-dehydrocholesterol) into vitamin D_3 (cholecalciferol). Vitamin D_3 (**iv**) within plasma combines with vitamin D-binding protein (DBP) and is first hydroxylated to 25 hydroxy-D_3 by liver enzymes. Final hydroxylation (**v**) of 25-hydroxy-D_3 takes place in the kidney forming 1, 25-dihydroxy vitamin D (calcitriol), biological active form of vitamin D, in the presence of cytochromic enzymes

Keeping in mind that several environmental triggers have plausible links to T1D, data from a number of large-scale studies are available to corroborate such claims. The Environmental Determinants of Diabetes in the Young (TEDDY), a prospective cohort study followed children during the first 15 years of their life. Various

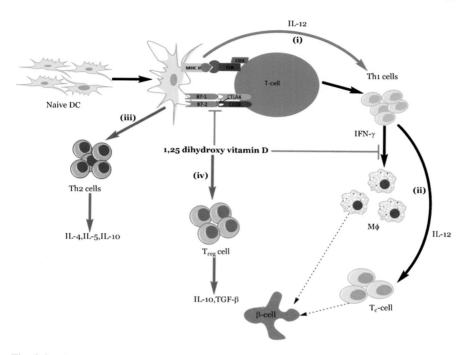

Fig. 2.4 Vitamin D is associated with various effects on the human body namely, maintaining adequate growth and calcium homeostasis. Vitamin D also affects the immune system and insulin levels, mediated via the VDR on the surface of immune and β-cell. (**i**) 1,25-dihydroxy vitamin D prevents the maturation of APCs, especially DC & Mφ, by inhibiting the expression of MHC II and adhesion molecules on the surface of DC's, predominantly by inhibiting the secretion of IL-12 leading to decreased Th1 cell stimulation. (**ii**) Th1 cell then leads into cytotoxic T-cell (T$_c$) release. (**iii**) Additionally, vitamin D induces Th2 cells, which in turn release anti-inflammatory cytokines such as IL-4, IL-5 and IL-10. Vitamin D further favors the production of T$_{reg}$ cells. Th2 and T$_{reg}$ cells collectively play a role in inhibiting Th1 cell population by producing counteracting cytokines

clinical tests were performed using blood, stool, and urine samples. The study aim was to identify a correlation among differing roles of infectious diseases, diet, toxins, etc. and T1D (Ziegler et al. 2011). Similarly, BABYDIAB was another prospective cohort study initiated in the late 1980s and concluded in 2009. The study included 1650 children born from diabetic parents, following their progress over a period of 20 years. Participants were tested for islet autoantibodies and the study aimed to identify genetic and environmental factors responsible for triggering the development of islet autoantibodies. The findings demonstrated that islet autoimmunity often initiates within the first 2 years of life although such prospective studies are limited in what they reveal and should ideally be extended beyond young adulthood; this could lead to the identification of major factors and facilitate appropriate primary prevention and intervention at different ages and stages of the disease (Hummel and Ziegler 2011).

References

Adams SF (1926) The seasonal variation in the onset of acute diabetes: the age and sex factors in 1,000 diabetic patients. Arch Intern Med 37(6):861–864

Åkerblom H (2011) The Trial to Reduce IDDM in the Genetically at Risk (TRIGR) study: recruitment, intervention and follow-up. Diabetologia 54(3):627–633

Åkerblom H, Virtanen S et al (2005) Dietary manipulation of beta cell autoimmunity in infants at increased risk of type 1 diabetes: a pilot study. Diabetologia 48(5):829–837

Åkerblom HK, Knip M (1998) Putative environmental factors in type 1 diabetes. Diabetes Metab Rev 14(1):31–68

Anjos S, Nguyen A et al (2002) A common autoimmunity predisposing signal peptide variant of the cytotoxic T-lymphocyte antigen 4 results in inefficient glycosylation of the susceptibility allele. J Biol Chem 277(48):46478–46486

Bell G, Horita S et al (1984) A polymorphic locus near the human insulin gene is associated with insulin-dependent diabetes mellitus. Diabetes 33(2):176–183

Borch-Johnsen K, Mandrup-Paulsen T et al (1984) Relation between breast-feeding and incidence rates of insulin—dependent diabetes mellitus: a hypothesis. Lancet 324(8411):1083–1086

Bottini N, Vang T et al (2006) Role of PTPN22 in type 1 diabetes and other autoimmune diseases. Semin Immunol 18(4):207–213

Bottini N, Musumeci L et al (2004) A functional variant of lymphoid tyrosine phosphatase is associated with type I diabetes. Nat Genet 36(4):337–338

Brekke HK, Ludvigsson J (2007) Vitamin D supplementation and diabetes-related autoimmunity in the ABIS study. Pediatr Diabetes 8(1):11–14

Buhlmann JE, Elkin SK et al (2003) A role for the B7-1/B7-2: CD28/CTLA-4 pathway during negative selection. J Immunol 170(11):5421–5428

Classen JB, Classen DC (1999) Immunization in the first month of life may explain decline in incidence of IDDM in The Netherlands. Autoimmunity 31(1):43–45

Classen JB, Classen DC (2002) Clustering of cases of insulin dependent diabetes (IDDM) occurring three years after hemophilus influenza B (HiB) immunization support causal relationship between immunization and IDDM. Autoimmunity 35(4):247–253

Coppieters KT, Boettler T et al (2012) Virus Infections in Type 1 Diabetes. Cold Spring Harb Perspect Med 2(1):a007682

Cooper JD, Smyth DJ et al (2011) Inherited variation in vitamin D genes is associated with predisposition to autoimmune disease type 1 diabetes. Diabetes 60(5):1624–1631

Danescu LG, Levy S et al (2009) Vitamin D and diabetes mellitus. Endocrine 35(1):11–17

Dawn TM, Barrett JH (2005) Genetic linkage studies. Lancet 366(9490):1036–1044

Davies JL, Kawaguchi Y et al (1994) A genome-wide search for human type 1 diabetes susceptibility genes. Nature 371(6493):130–136

Elliott R, Martin J (1984) Dietary protein: a trigger of insulin-dependent diabetes in the BB rat? Diabetologia 26(4):297–299

Espino-Paisan L, De La Calle H et al (2011) Study of polymorphisms in 4q27, 10p15, and 22q13 regions in autoantibodies stratified type 1 diabetes patients. Autoimmunity 44(8):624–630

Füchtenbusch M, Irnstetter A et al (2001) No evidence for an association of coxsackie virus infections during pregnancy and early childhood with development of islet autoantibodies in offspring of mothers or fathers with type 1 diabetes. J Autoimmun 17(4):333–340

Fujinami RS, Von Herrath MG et al (2006) Molecular mimicry, bystander activation, or viral persistence: infections and autoimmune disease. Clin Microbiol Rev 19(1):80–94

Gamble DR, Kinsley ML et al (1969) Viral antibodies in diabetes mellitus. Br Med J 3(5671):627–630

Garg G, Tyler JR et al (2012) Type 1 diabetes-associated IL2RA variation lowers IL-2 signaling and contributes to diminished $CD4^+$ $CD25^+$ regulatory T cell function. J Immunol 188 (9):4644–4653

Gillespie KM, Boraska V (2011) The genetics of type 1 diabetes. In: Wagner D (ed) type 1 diabetes—pathogenesis, genetics and immunotherapy. ISBN: 978-953-307-362-0, InTech, DOI: 10.5772/21924. Available from: http://www.intechopen.com/books/type-1-diabetes-pathogenesis-genetics-and-immunotherapy/the-genetics-of-type-1-diabetes

Graves PM, Rotbart HA et al (2003) Prospective study of enteroviral infections and development of beta-cell autoimmunity: diabetes autoimmunity study in the young (DAISY). Diabetes Res Clin Pract 59(1):51–61

Group (1999) Vitamin D supplement in early childhood and risk for Type I (insulin-dependent) diabetes mellitus. Diabetologia 42(1):51–54

Helgason T, Ewen S et al (1982) Diabetes produced in mice by smoked/cured mutton. Lancet 320 (8306):1017–1022

Helgason T, Jonasson MR (1981) Evidence for a food additive as a cause of ketosis-prone diabetes. Lancet 318(8249):716–720

Hewison M, Freeman L et al (2003) Differential regulation of vitamin D receptor and its ligand in human monocyte-derived dendritic cells. J Immunol 170(11):5382–5390

Hirschhorn JN, Daly MJ (2005) Genome-wide association studies for common diseases and complex traits. Nat Rev Genet 6(2):95–108

Hou J, Said C et al (1994) Antibodies to glutamic acid decarboxylase and P2-C peptides in sera from coxsackie virus B4-infected mice and IDDM patients. Diabetes 43(10):1260–1266

Hummel S, Ziegler AG (2011) Early determinants of type 1 diabetes: experience from the BABYDIAB and BABYDIET studies. Am J Clin Nutr 94(6 suppl):1821S–1823S

Jun HS, Yoon JW (2003) A new look at viruses in type 1 diabetes. Diabetes Metab Res Rev 19 (1):8–31

Karvonen M (2006) Incidence and trends of childhood type 1 diabetes worldwide 1990–1999. Diabet Med 23(8):857–866

Kennedy GC, German MS et al (1995) The minisatellite in the diabetes susceptibility locus IDDM2 regulates insulin transcription. Nat Genet 9(3):293–298

Kimpimäki T, Erkkola M et al (2001) Short-term exclusive breastfeeding predisposes young children with increased genetic risk of type I diabetes to progressive beta-cell autoimmunity. Diabetologia 44(1):63–69

Knip M, Åkerblom HK et al (2014) Hydrolysed infant formula and early β-cell autoimmunity: a randomized clinical trial. JAMA 311(22):2279–2287

Knip M, Virtanen SM et al (2010) Dietary intervention in infancy and later signs of beta-cell autoimmunity. N Engl J Med 363(20):1900–1908

Knip M, Veijola R et al (2005) Environmental triggers and determinants of type 1 diabetes. Diabetes 54(Suppl 2):S125–S136

Kriegel MA, Manson JAE et al (2010) Does vitamin D affect risk of developing autoimmune disease?: a systematic review. Semin Arthritis Rheum 40(6):512–531

Lyons PA, Wicker LS (1999) Localising quantitative trait loci in the NOD mouse model of type 1 diabetes. Curr Dir Autoimmun 1:208–225

Mathieu C, Badenhoop K (2005) Vitamin D and type 1 diabetes mellitus: state of the art. Trends Endocrinol Metab 16(6):261–266

Mathieu C, Laureys J et al (1992) 1,25-Dihydroxyvitamin D3 prevents insulitis in NOD mice. Diabetes 41(11):1491–1495

National human genome research institute. National Institutes of Health. Genome.gov. Genome-Wide Association Studies: fact sheets. Last update January 2014

Nielsen DS, Krych Ł et al (2014) Beyond genetics. Influence of dietary factors and gut microbiota on type 1 diabetes. FEBS Letters pii:S0014-5793(14)00296-8

Nielsen L, Pörksen S et al (2011) The PTPN22 C1858T gene variant is associated with proinsulin in new-onset type 1 diabetes. BMC Med Genet 12(1):41

Norris JM, Barriga K et al (2005) Risk of celiac disease autoimmunity and timing of gluten introduction in the diet of infants at increased risk of disease. JAMA 293(19):2343–2351

Oikarinen M, Tauriainen S et al (2008) Detection of enteroviruses in the intestine of type 1 diabetic patients. Clin Exp Immunol 151(1):71–75

Peechakara SV, Pittas AJ (2008) Vitamin D as a potential modifier of diabetes risk. Nat Clin Pract Endocrinol Metab 4(4):182–183

Pociot F, McDermott MF (2002) Genetics of Type 1 diabetes mellitus. Genes Immun 3(5):235–249

Pociot F, Akolkar B et al (2010) Genetics of type 1 diabetes: what's next? Diabetes 59(7):1561–1571

Pugliese A, Zeller M et al (1997) The insulin gene is transcribed in the human thymus and transcription levels correlate with allelic variation at the INS VNTR-IDDM2 susceptibility locus for type 1 diabetes. Nat Genet 15(3):293–297

Pugliese A, Gianani R et al (1995) HLA-DQB1*0602 is associated with dominant protection from diabetes even among islet cell antibody-positive first-degree relatives of patients with IDDM. Diabetes 44(6):608–613

Risch N (1987) Assessing the role of HLA-linked and unlinked determinants of disease. Am J Hum Genet 40(1):1–14

Salminen K, Sadeharju K et al (2003) Enterovirus infections are associated with the induction of β-cell autoimmunity in a prospective birth cohort study. J Med Virol 69(1):91–98

Salomon B, Bluestone JA (2001) Complexities of CD28/B7: CTLA-4 costimulatory pathways in autoimmunity and transplantation. Ann Rev Immunol 19(1):225–252

Schloot N, Willemen S et al (2001) Molecular mimicry in type 1 diabetes mellitus revisited: T-cell clones to GAD65 peptides with sequence homology to Coxsackie or proinsulin peptides do not crossreact with homologous counterpart. Hum Immunol 62(4):299–309

Stene L, Ulriksen J et al (2001) Use of cod liver oil during pregnancy associated with lower risk of type I diabetes in the offspring. Diabetologia 43(9):1093–1098

Tauriainen S, Oikarinen S et al (2011) Enteroviruses in the pathogenesis of type 1 diabetes. Semin Immunopathol 33(1):45–55

Todd JA, Bell JI et al (1987) HLA-DQB gene contributes to susceptibility and resistance to insulin-dependent diabetes mellitus. Nature 329(6140):599–604

Tull E, Tajima N et al (1992) Epidemics, migrants and the death of the pancreas. In: Levy-Marchal C, Czernichow P (ed) Pediatric and adolescent endocrinology. Epidemiology and etiology of insulin-dependent diabetes in the young, vol 21. Karger, pp 56–65

Vafiadis P, Bennett ST et al (1997) Insulin expression in human thymus is modulated by INS VNTR alleles at the IDDM2 locus. Nat Genet 15(3):289–292

Virtanen S, Jaakkola L et al (1994) Nitrate and nitrite intake and the risk for type 1 diabetes in Finnish children. Diabet Med 11(7):656–662

Viskari H, Ludvigsson J et al (2005) Relationship between the incidence of type 1 diabetes and maternal enterovirus antibodies: time trends and geographical variation. Diabetologia 48 (7):1280–1287

Viskari H, Koskela P et al (2000) Can enterovirus infections explain the increasing incidence of type 1 diabetes? Diabetes Care 23(3):414–416

Wang CJ, Schmidt EM et al (2011) Immune regulation by CTLA-4—relevance to autoimmune diabetes in a transgenic mouse model. Diabetes Metab Res Rev 27(8):946–950

Wang XB, Zhao X et al (2002) A CTLA4 gene polymorphism at position -318 in the promoter region affects the expression of protein. Genes Immun 3(4):233–234

Wiebolt J, Koeleman BPC et al (2010) Endocrine autoimmune disease: genetics become complex. Eur J Clin Invest 40(12):1144–1155

Zeitz U, Weber K et al (2003) Impaired insulin secretory capacity in mice lacking a functional vitamin D receptor. FASEB J 17(3):509–511

Ziegler AG, Pflueger M et al (2011) Accelerated progression from islet autoantibody to diabetes is causing the escalating incidence of type 1 diabetes in young children. J Autoimmun 37(1):3–7

Ziegler AG, Schmid S et al (2003) Early infant feeding and risk of developing type 1 diabetes–associated autoantibodies. JAMA 290(13):1721–1728

Zipitis CS, Akobeng AK (2008) Vitamin D supplementation in early childhood and risk of type 1 diabetes: a systematic review and meta-analysis. Arch Dis Child 93(6):512–517

Chapter 3
Predictors and Pathogenesis of Type 1 Diabetes

3.1 Key Predictors of Type 1 Diabetes

Autoimmune processes in general take many years before their clinical manifestation as a disease becomes apparent and T1D is no exception, sometimes taking up to a decade, or more for clinical presentation. This asymptomatic period offers a great window of opportunity for the prediction and prevention of full-blown disease (Fig. 3.1). Different groups worldwide have investigated and found several T1D age-associated biomarkers, these being genetic, immunological, and metabolic in nature. These biomarkers facilitate tracking of the immunological checkpoint failure leading to progression toward clinical T1D (Fig. 3.2a). Prior to embark on an in-depth discussion related to the host of biomarkers, some caution is necessary. Our current knowledge is primarily dependent on several assumptions and data extrapolation, and hence caveats are inevitable. Allied glitches include that the following: (1) most of the data are from studies on hereditary T1D cohorts, which present a minor portion of the entire T1D population. (2) The study groups primarily include children and young adults while there is also a noteworthy proportion of T1D in the population who develop autoimmunity in late adulthood. (3) Despite the presence of range of reliable genetic and immunological biomarkers, there remains a dearth of biomarkers specifically related to $CD4^+$ and $CD8^+$ effector and T_{reg} cells, as direct evidence of pathogenesis and T1D (Cassiday 2008; Ziegler and Nepom 2010).

As described above, in addition to genetic susceptibility, various other trigger agents are involved in the initiation and progression of autoimmune T1D. For example, the presence of autoantibodies is often observed years before the clinical onset of T1D (Leslie et al. 2001). Reports suggest that autoantibodies may be transferred via the placenta or develop after birth, but this is rarely diagnosed before

© The Author(s) 2016
P. Singh et al., *Therapeutic Perspectives in Type-1 Diabetes*,
SpringerBriefs in Applied Sciences and Technology,
DOI 10.1007/978-981-10-0602-9_3

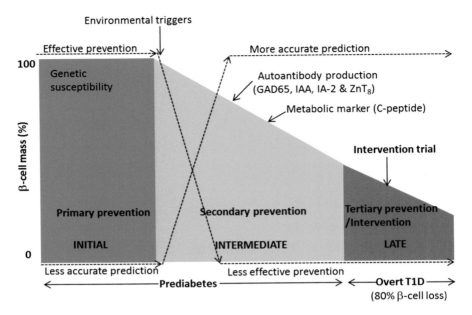

Fig. 3.1 Model showing the progress of disease and potential stages of predictability and prevention of T1D relating it to the loss of β-cell mass and time. Modified from Eisenbarth (1986)

the age of 6 months (Rubio-Cabezas et al. 2009). More than two dozen autoantigen-related autoantibodies have been implicated although four major islet autoantibodies have drawn significant attention: insulin (or proinsulin or IAA) (Palmer et al. 1983), GAD isoforms (65 and 67) (Baekkeskov et al. 1990), tyrosine phosphatase-related islet antigens (IA-2 or ICA512) (Rabin et al. 1994), and the most recently identified zinc transporter (ZnT8) (Wenzlau et al. 2007). However, the positive detection of any one of the autoantibodies mentioned only imparts a marginal increase in the risk of developing T1D and to date it has not been possible to reliably predict progression to clinical disease although incremental increases in disease risk were found in individuals who were positive for two, three, or more autoantibodies (Fig. 3.2b) (Verge et al. 1996). Relative risk was also found to vary dependant on autoantibody type with some reports emerging that the most influencing antibody is insulin, as identified in NOD mice (where diabetes was inhibited by single amino acid mutation in insulin peptide 9-23) and as the only β-cell-specific autoantibody identified in T1D (Nakayama et al. 2005). In contrast, several other studies support GAD65 and IA-2 as high risk-associated autoantibodies in humans. Nonetheless, higher numbers of autoantibodies, high titre, and affinity for epitopes to various antigens govern the degree of risk associated to any particular antigen group (Verge et al. 1998).

In the late 1990s, when screening using recombinant antigen assay methods were common place, about 5–10 % of the adult T2D population had autoantibodies

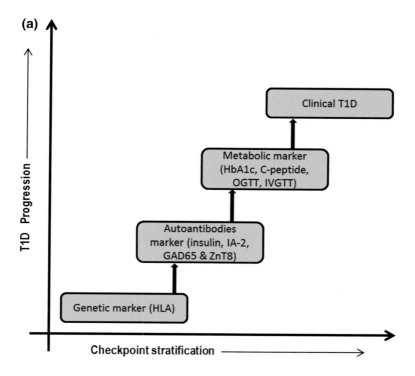

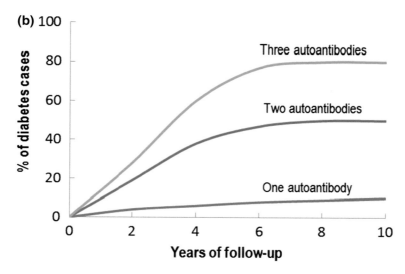

Fig. 3.2 a Checkpoint classification based on the biomarkers used to indicate T1D progression, **b** Graphical representation correlating autoantibody numbers and % cases of first-degree relatives with T1D progression based on several longitudinal studies (OGTT = Oral glucose tolerance test; IVGTT = Intravenous glucose tolerance test; HbA1C = Glycated hemoglobin). Modified from Notkins and Lernmark (2001)

for GAD65 and a slightly lower percentage (2–4 %), tested positive for IA-2. If this is the case, it could be inferred that many of the patients who were classified as T2D sufferers, actually either had T1D or a combination of T1D and T2D (Turner et al. 1997; Hawa et al. 2000). Based on this hypothesis, it could further be concluded that the number of T1D patients would actually be double than the current estimates. The current approach to diagnosis includes a stepwise decision tree to identify individuals at high risk of developing T1D (see Fig. 3.2a). Initially, genotyping for susceptible HLA II genes and patients with a family history is tested. Individuals or children (≥age 1 year) who warrant sufficient HLA susceptibility in the first stage are then measured for their autoantibody levels. Additional genotyping in autoantibody positive individuals helps exclude the one with protective HLA II genes. As a final check, often insulin-secreting capacity is measured as an indicator of β-cell function or IVGTT, OGTT, and C-peptide levels are also measured (Srikanta et al. 1985).

Cumulative evidence indicates the multifactorial nature of T1D due to associated complex immunology, and this necessitates the use of a combinatorial screening method to identify the individuals at risk of developing T1D. In addition, continual efforts toward extracting reliable, specific biomarkers have led to the identification of some promising serum proteins, namely cytokines and chemokines (Wasserfall and Atkinson 2006). Cytokines have been proposed as one of the major inducers of β-cell damage especially the type 1 cytokines (Th1) such as interferon gamma (IFN-γ), tumor necrosis factor-α (TNF-α), and IL-2 and they are considered 'high risk' with approximately 40 % of patients going on to develop T1D within 5 years (Ryden et al. 2009). However, a lack of specificity of cytokines for T1D, their varied level of expression, and trace amount poses serious analytical challenges for their quantification, when placed against a background of large numbers of non-relevant proteins (Borges et al. 2010; Carey et al. 2010). Despite considerable global efforts and investment progress in the area of T1D disease prediction has progressed at a painfully slow pace, with researchers keenly anticipating that advancements in proteomics and metabolomics will serve as a future platform for the identification of reliable, specific biomarkers for T1D in the coming era (Cassiday 2008).

3.1.1 Pathophysiology of Type 1 Diabetes: What Goes Awry?

3.1.1.1 T-Cells: A Double Edge Sword

Given that T1D is primarily an autoimmune disease, the mechanisms of which are poorly understood, the actions of a handful of antigens such as insulin, GAD isoforms (GAD65 & 67), IA-2, and ZnT8 are known to transform into autoantigens. In individuals of high susceptibility the pancreatic Islet of Langerhans are primarily infiltrated by APCs such as DCs and Mφs. Concomitantly or shortly thereafter autoreactive CD4$^+$ and CD8$^+$ T-cells can be readily detected surrounding the islets (known as peri-insulitis, see Fig. 3.3) (Turley et al. 2003) as highlighted in a study by Haskins et al., where it was demonstrated that progression of autoimmune diabetes occurred in NOD mice that received adoptive transfer of CD4$^+$-specific T-cells from NOD mice compared to placebo (Haskins and McDuffie 1990;

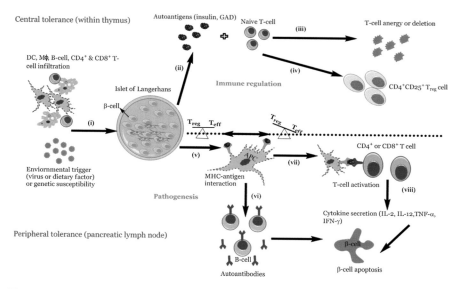

Fig. 3.3 Pathogenetic model of T1D. In individuals of high susceptibility, the pancreatic Islet of Langerhans are primarily infiltrated by APCs such as DCs and Mφs followed by autoreactive CD4$^+$ and CD8$^+$ T-cells which can be detected surrounding the islets (known as peri-insulitis) (**i**). This islet damage is followed by autoantigen release (**ii**) and while on one hand thymic immune regulation leads to active tolerance by deleting T-cells (**iii**) or inducing protective, regulatory T-cells (CD4$^+$ CD25$^+$ T$_{reg}$) (**iv**) on other hand its failure leads to escape of autoreactive T-cells into peripheral pancreatic lymph nodes, inducing an autoimmune response. In order to prevent this autoimmune response it is essential to maintain a balance between the level of T$_{reg}$ ('good cells') and effector T$_{eff}$ ('bad cells') T-cells. The released autoantigens are taken up by the APCs (mainly DCs) (**v**) and from here, there is a diversion in the fate of APCs. One path leads to APCs stimulating B-lymphocytes, which produces autoantibodies toward the autoantigens (**vi**) (Lennon et al. 2009) and the role of these B-lymphocytes is briefly discussed below. Another possibility is the conversion of the autoantigens into peptides followed by pMHC–peptide complex formation. Further, these complexes are presented onto CD4$^+$ or CD8$^+$ (T$_{eff}$ cells) TCRs which activate T-cells (**vii**). This causes the release of proinflammatory cytokines and chemokines (IL-2, TNF-α, and CTL) (**viii**) and ultimately leads to β-cell apoptosis resulting in eventual loss of insulin production once a critical mass of β-cells have been eliminated (Ounissi-Benkalha and Polychronakos 2008; Atkinson et al. 2011)

Wong et al. 1996). Due to ethical concerns, only recently the examination of pancreatic biopsy in prediabetic patients been possible; this makes its feasible to study the composition of any lesions (insulitis).

3.1.1.2 B-Lymphocytes Producing Diabetes-Associated Autantibodies

Autoantibodies against the generated autoantigens described above are often produced during the prodromal stages of T1D development. Although currently visualized as gold standard markers in T1D prognosis, it is mostly believed that generation of these autoantibodies by B-lymphocytes is rather a secondary consequence of ongoing β-cell destruction. Ample evidence exists which indicates that autoreactive T-cells are the primary arbitrator in T1D pathogenesis; however, there is also some circumstantial evidence on the role that B-cells play in T1D pathogenesis.

The notion of B-cell involvement in diabetes etiology is primarily supported by data obtained from studies conducted in NOD mice. Strong resistance to developing T1D was observed in NOD mice which were made deficient in B-cells through introduction of a functionally inactivated immunoglobulin μ heavy chain gene (NOD.Igμ^{null}). Reintroduction of B-cells in NOD.Igμ^{null} mice nullified their T1D resistance effect (Akashi et al. 1997; Wong et al. 1998). Taken together, this data begins to support the idea that in NOD mice, B-cells play a vital role in disease pathogenesis. However, in contrast to results reported from studies conducted in NOD mice, the implications of B-cell activity in T1D in human remain unclear.

The uncertainties lead us to subsequently question why and precisely when these protective B-cells lose their tolerance leading them to exert pathogenic effector functions. Explanations include their role as an APC and as discussed earlier, MHC haplotypes present a primary risk factor for the development of autoimmune T1D in humans and NOD mice (Lyons and Wicker 1999). MHC class II molecules are mostly expressed on hematopoietically derived APCs such as Mφ, B-cells, and DCs. Nevertheless, normal regulatory responses of APCs include binding and presentation of antigenic peptide or autoantigens to class I or class II MHC molecules in a tolerogenic manner. However, impaired presentation of peptide MHC molecule leads to the activation of autoreactive CD4$^+$ and CD8$^+$ T-cells (Sebzda et al. 1999; Moser 2003). Due to the presence of specific Ig antibodies on the B-cell surface, they are efficient in capture and internalization of antigens and in presenting them to CD4$^+$ T-cells. Another scenario also detailed earlier is that B-cells can go on to produce islet-specific autoantibodies against β-cell antigens and in view of this it would not be erroneous to conclude that, despite the fact that autoantibodies are the best available diagnostic tools for T1D prediction, their ability to modulate pancreatic β-cell damage in vivo remains controversial; on the other hand, the success with anti-CD3 and anti-CD20 antibodies has suggested a preventive role of B-cell therapy in T1D, as is discussed below.

References

Akashi T, Nagafuchi S et al (1997) Direct evidence for the contribution of B cells to the progression of insulitis and the development of diabetes in non-diabetic mice. Int Immunol 9:1159–1164

Atkinson MA, Bluestone JA et al (2011) How does Type 1 diabetes develop?: the notion of homicide or β-cell suicide revisited. Diabetes 60(5):1370–1379

Baekkeskov S, Aanstoot HJ et al (1990) Identification of the 64 K autoantigen in insulin-dependent diabetes as the GABA-synthesizing enzyme glutamic acid decarboxylase. Nature 347(6289):151–156

Borges CR, Rehder DS et al (2010) Full-length characterization of proteins in human populations. Clin Chem 56(2):202–211

Carey C, Purohit S et al (2010) Advances and challenges in biomarker development for type 1 diabetes prediction and prevention using 'omic' technologies. Expert Opin Med Diagn 4(5):397–410

Cassiday L (2008) Candidate biomarkers for type 1 diabetes. J Proteome Res 7(2):482

Eisenbarth G (1986) Type I diabetes mellitus. A chronic autoimmune disease. N Engl J Med 314(21):1360–1368

Haskins K, McDuffie M (1990) Acceleration of diabetes in young NOD mice with a CD4$^+$ islet-specific T cell clone. Science 249(4975):1433–1436

Hawa MI, Fava D et al (2000) Antibodies to IA-2 and GAD65 in type 1 and type 2 diabetes: isotype restriction and polyclonality. Diabetes Care 23(2):228–233

Lennon GP, Bettini M et al (2009) T cell islet accumulation in type 1 diabetes is a tightly regulated, cell-autonomous event. Immunity 31(4):643–653

Leslie D, Lipsky P et al (2001) Autoantibodies as predictors of disease. J Clin Invest 108(10):1417–1422

Lyons PA, Wicker LS (1999) Localising quantitative trait loci in the NOD mouse model of type 1 diabetes. Curr Dir Autoimmun 1:208–225

Moser M (2003) Dendritic cells in immunity and tolerance-do they display opposite functions? Immunity 19:5–8

Nakayama M, Abiru N et al (2005) Prime role for an insulin epitope in the development of type 1 diabetes in NOD mice. Nature 435(7039):220–223

Notkins AL, Lernmark A (2001) Autoimmune type 1 diabetes: resolved and unresolved issues. J Clin Invest 108(9):1247–1252

Ounissi-Benkalha H, Polychronakos C (2008) The molecular genetics of type 1 diabetes: new genes and emerging mechanisms. Trends Mol Med 14(6):268–275

Palmer JP, Asplin CM et al (1983) Insulin antibodies in insulin-dependent diabetics before insulin treatment. Science 222(4630):1337–1339

Rabin DU, Pleasic SM et al (1994) Islet cell antigen 512 is a diabetes-specific islet autoantigen related to protein tyrosine phosphatases. J Immunol 152(6):3183–3188

Rubio-Cabezas O, Minton JAL et al (2009) Clinical heterogeneity in patients with FOXP3 mutations presenting with permanent neonatal diabetes. Diabetes Care 32(1):111–116

Ryden A, Stechova K et al (2009) Switch from a dominant Th1-associated immune profile during the pre-diabetic phase in favour of a temporary increase of a Th3-associated and inflammatory immune profile at the onset of type 1 diabetes. Diabetes Metab Res Rev 25(4):335–343

Sebzda E, Mariathasan S et al (1999) Selection of the T cell repertoire. Annu Rev Immunol 17(1):829–874

Srikanta S, Ganda OP et al (1985) First-degree relatives of patients with type I diabetes mellitus. islet-cell antibodies and abnormal insulin secretion. N Eng J Med 313(8):461–464

Turley S, Poirot L et al (2003) Physiological β cell death triggers priming of self-reactive T cells by dendritic cells in a type-1 diabetes model. J Exp Med 198(10):1527–1537

Turner R, Stratton I et al (1997) UKPDS 25: autoantibodies to islet-cell cytoplasm and glutamic acid decarboxylase for prediction of insulin requirement in type 2 diabetes. The Lancet 350(9087):1288–1293

Verge CF, Stenger D et al (1998) Combined use of autoantibodies (IA-2 autoantibody, GAD autoantibody, insulin autoantibody, cytoplasmic islet cell antibodies) in type 1 diabetes: combinatorial islet autoantibody workshop. Diabetes 47(12):1857–1866

Verge CF, Gianani R et al (1996) Prediction of type I diabetes in first-degree relatives using a combination of insulin, GAD, and ICA512bdc/IA-2 autoantibodies. Diabetes 45(7):926–933

Wasserfall CH, Atkinson MA (2006) Autoantibody markers for the diagnosis and prediction of type 1 diabetes. Autoimmun Rev 5(6):424–428

Wenzlau JM, Juhl K et al (2007) The cation efflux transporter ZnT8 (Slc30A8) is a major autoantigen in human type 1 diabetes. Proc Natl Acad Sci U S A 104(43):17040–17045

Wong FS, Visintin I et al (1998) The role of lymphocyte subsets in accelerated diabetes in nonobese diabetic-rat insulin promoter–B7-1 (NOD-RIP-B7-1) mice. J Exp Med 187(12): 1985–1993

Wong FS, Visintin I et al (1996) CD8 T cell clones from young nonobese diabetic (NOD) islets can transfer rapid onset of diabetes in NOD mice in the absence of CD4 cells. J Exp Med 183(1):67–76

Ziegler AG, Nepom GT (2010) Prediction and pathogenesis in type 1 diabetes. Immunity 32(4):468–478

Chapter 4
Type 1 Diabetes: Past, Present, and Future Therapies

4.1 Introduction

The first line of treatment developed for T1D was insulin. Over the last century, several new approaches to treat T1D have been adopted (Fig. 4.1), but insulin delivery still remains the main stream of treatment. The new strategies to treat T1D include replacement of β-cell via transplantation or regeneration, but are still in their infancy. Another vibrant approach in T1D immunotherapy involves compensating for the lost pancreatic β-cells and restoration of immunological tolerance by islet like cells derived from stem cells. The major benefit associated with these intervention strategies is the ease of patient identification and efficacy evaluation within a much shorter duration. Earlier reports on NOD mice had provided the basis for antigenic and non-antigenic preventive strategies to treat T1D, but a number of clinical trials have been completed with limited success so far. An ideal intervention for T1D would require an entity to halt autoimmune response along with an additional element to enhance β-cell function or expedite its regeneration. As such, combination therapy is believed to lead the treatment of T1D in the time to come.

4.2 Insulin Replacement

Discovery and isolation of insulin in the year 1922 by Best and Banting provided the first ray of hope to the many thousands of diabetic patients of the time (Banting and Best 1922). Albeit the fact that enormous resources have since been directed toward finding a cure for T1D, insulin replacement remains the mainstay of symptomatic treatment today. Conventional insulin delivery methods include (1) subcutaneous (sc), which is highly encouraged due to low cost and ease of delivery, (2) oral administration, and (3) intramuscular (im) injection, both of which are less popular due to partial insulin loss in the gastrointestinal tract in the former

© The Author(s) 2016
P. Singh et al., *Therapeutic Perspectives in Type-1 Diabetes*,
SpringerBriefs in Applied Sciences and Technology,
DOI 10.1007/978-981-10-0602-9_4

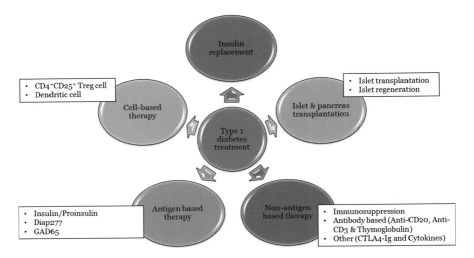

Fig. 4.1 Current therapeutic intervention strategies for T1D

and high dose requirement in the latter case, respectively. Nonetheless, several drawbacks have raised concerns on the use of insulin, including the need for effective glucose level monitoring, which otherwise lead to hyper- or hypogly-caemia and, in chronic cases of mismanagement, severe microvascular complica-tions. Additionally, patients have a life-long dependency on insulin (Fujinami et al. 2006). Several state-of-the-art approaches are being developed and deployed to support ease of insulin administration such as the introduction of insulin pumps, inhalers, jet injectors, transdermal patches, and orally administered formulations. Insulin pumps, inhalers, and injectors are currently in the market and have drawn considerable appreciation of both patients and clinicians; alike, their widespread usage has been hampered by high costs and not user-friendly devices (Haller et al. 2005). Key challenges pertaining to the efficacious use of insulin depend not only on ensuring their effective delivery system but also on the correct dose, time, and frequency of delivery.

Further progress in the field of type 1 diabetes patients has given rise to real-time continuous glucose monitoring (CGM). One of the first marketed product based on this technology was the GlucoWatch Biographer® (Cygnus Inc., Redwood City, CA) currently owned by Animas Corporation (Johnson & Johnson Companies, Milpitas, CA). Numerous other approved CGM devices include DexCom Seven® (DexCom, San Diego, CA), Paradigm RT®, and Freestyle Navigator (Abbott Diabetes Care, Alameda, California). These CGM devices are considered an adjuvant to self-monitoring of blood glucose in order to confirm information, such as hyper- and hypo-glycaemic events (Ellis et al. 2008; Renard 2002). CGMs with continuous subcutaneous insulin infusion (CSII) promoted research toward closed-loop systems. This system allows the delivery of insulin to the changing glucose levels in a real-time response. A closed-loop approach holds revolutionary

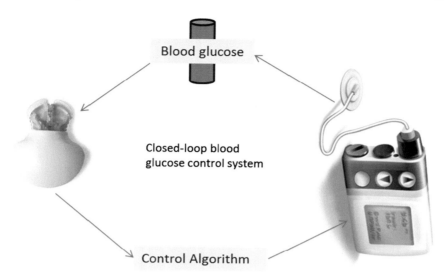

Fig. 4.2 Key elements involved in the restoration of closed-loop blood glucose control system in type 1 diabetes

power for glucose management. A closed-loop system, otherwise known as the artificial pancreas, comprises three parts, a CGM device to monitor glucose concentration, a titration algorithm to calculate the amount of insulin to be delivered, and an insulin pump to deliver the computed dosage of insulin (Fig. 4.2). Primarily, three types of closed-loop systems are defined as follows: (1) a sc–sc system with a subcutaneous (sc) glucose sensing and sc insulin delivery, (2) an iv–ip with an iv sensing and intraperitonial (ip) delivery system, and (3) an iv–iv system with iv sensing and iv delivery (Renard 2015; Renard and Schaepelynck 2007). Given the minimal invasion and pervasive usage of external insulin pumps, the sc–sc approach had the potential to achieve a widespread application. Even so it may not be compatible with a fully closed-loop system due to considerable delays prohibiting safe and timely control of large or rapidly absorbed meals. Iv–iv approach is used in particular situations such as in the critically ill patients, comatose, or for research investigations. The shortcoming of this approach is its invasiveness which jeopardizes patient to infections and clotting problems (Hovorka 2008). Therefore, the uses of closed-loop systems are particularly appealing, and these include glucose monitoring with an insulin pump that is encompassed within an artificial pancreas (Powers 2008; Hovorka 2005). These devices not only closely monitor the blood glucose level but also deliver insulin in body in a controlled way. They aim to reduce the risk of hypoglycaemia while achieving tight control of glucose (Elleri et al. 2011). Clinical testing on closed-loop systems so far presented positive results with regards to safety and efficacy. It was found that overnight closed-loop delivery of insulin improved overnight glycaemic control and less risk of nocturnal hypoglycaemia compared to patients treated with conventional insulin pump (Breton et al. 2012; Buckingham et al. 2010). One of the novel techniques for continuous immunosensors is weak affinity chromatography which can sense molecules with fast off-rate and disassociation constant (K_d) in $\approx$ mM range necessary for detection of dynamic change in analyte concentration in the blood. Monoclonal Ab (3F1E8-A2) has been evaluated for their future potential in glucose sensor and was found to be specific for glucose molecule compared to other monosaccharides with a Kd = 18.8 ± 2.6 mM (Engström et al. 2008). The major stepping stone in the glucose sensor future will be its clinical validation for diabetes control.

4.3 β-Cell Replacement: Transplantation Versus Regeneration

4.3.1 Islet and Pancreas Transplantation

An alternative therapeutic option to conventional insulin replacement therapy is based on islet or pancreatic transplantation and replacement of β-cells with insulin-producing cells obtained from sources other than islet cells. The objective behind replacement of β-cells either with whole pancreas or islet transplantation is to attain long-term insulin independence, and the associated benefits which include better quality of life and prevention of secondary complications such as nephropathy, retinopathy, and vasculopathy (Sutherland et al. 2004). Pancreatic transplantation is an established procedure using the Edmonton protocol (Shapiro et al. 2006). Primary results revealed poor graft survival rates of less than 5 % after 6 months (Kelly et al. 1968) although simultaneous transplantation of a whole pancreas and kidney, albeit highly invasive, has delivered significant improvements in patient response and prognosis (Humar et al. 2000).

   Comparatively, islet transplantation seems a more promising approach for β-cell replacement since it is a noninvasive procedure that involves percutaneous cannulation of the portal vein. Furthermore, islet grafts can be stored in culture tissue or can be cryopreserved (Ryan et al. 2001; Nanji and Shapiro 2006). Despite advancements in this field, a 5-year follow-up study showed that a mere 10 % of the tested population remained insulin independent for a median period of 15 months after transplantation (Ryan et al. 2005). In fact, a major setback to the approach is the difficulty to protect the transplanted islets from host immune attack (Humar et al. 2000).

   However, advancements in the techniques to procure islets from cadaveric donors, less toxic immunosuppression regimens, and techniques such as xenotransplantation which makes use of islets from sources other than humans (e.g., porcine) will no doubt provide the much required edge to islet transplantation as a choice of therapy in the imminent future. Immune reactions against xenotransplant tissue are one of the lingering hurdles associated with this approach. Nevertheless, immunosuppression alongside innovative techniques such as islet encapsulation in alginate microcapsules, which protect the islets from T-cell mediated responses, have shown much promise to provide breakthroughs leading to greater immune tolerance of xenotransplanted tissues (Zimmermann et al. 2005). In addition to immune tolerance following xenotransplantation, there also exists a very real risk of porcine endogenous retrovirus transmission to the host (Limbert et al. 2008) and hitherto data available to establish the prevalence of infection in transplanted patients remain inconclusive, questioning the widespread use of this approach.

4.3.2 Islet Regeneration

As described above, development of T1D occurs as a consequence of loss of functional β-cells. With this in mind, the primary goal of therapeutic interventions to treat T1D is to restore β-cell mass, in sufficient amount to maintain euglycemia. Encouraging results with clinical islet transplantation trials bolstered a novel idea to exploit the regeneration capacity of the pancreas and other related tissues (SoRelle and Naziruddin 2011). It is well established that in healthy individuals, lost β-cells are constantly replenished in vivo through β-cell regeneration; however, further progress in this area has been delayed due to a poor understanding of the mechanisms driving the replenishment process. Some scientists argue that it happens via *β-cell* neogenesis while, according to other experts, *self-replication* of β-cells takes place. As evidence, postnatal origin of β-cells using a genetic lineage method in pancreatectomized mice was investigated. The study demonstrated that mature β-cells possess the capacity to proliferate; the replacement of lost β-cells is mainly due to replication of the existing cell mass (Dor et al. 2004), whereas β-cell neogenesis occurs through progenitor cell activation and/or transdifferentiation of mature fully differentiated cells, as a means of β-cell mass expansion (Lipsett et al. 2006).

Current avenues under investigation for islet cell expansion include ex vivo expansion, which require the addition of pharmacological agents, and encouraging results have been reported in both in vivo animal and in vitro human models (Beattie et al. 1997). Various hormones and growth factors are also known to increase β-cell mass; these include gastrin, glucagon-like peptide (GLP-1), epidermal growth factor (EGF), and islet neogenesis-associated protein (INGAP). Regrettably, over the course of time these differentiated cells lose their capacity to secrete insulin, possibly due to aging. Till date, several combination therapies with EGF + gastrin and GLP-1 + gastrin have shown promising results in β-cell expansion (Brand et al. 2002; Suarez-Pinzon et al. 2008). Unfortunately, none of the studies have focused its application specifically on T1D.

Another pharmaceutically synthesized ligand under investigation is INGAP (INGAP[104–118]), first identified in hamster model of islet expansion. INGAP administration was found to increase functional β-cell mass in animals and therapeutically reverse hyperglycemia in streptozotocin-induced (STZ) diabetic models. The authors demonstrated that this pentadecapeptide is capable of inducing duct and acinar cell-associated islet neogenesis, which in turn was associated with

PDX-1 expression. PDX-1 is considered an important transcription factor required in normal islet development (Rosenberg et al. 2004; Jamal et al. 2005).

4.3.3   Stem Cell-Derived β-Cell Replacement

In order to address critical shortfalls of islet donors, stem cells appear to be an excellent alternative, being an unlimited source of β-cells, with the added potential of high differentiation and tissue regeneration. The most commonly used type of stem cells are embryonic (ESC) and autologous adult stem cells (ASC). ESC are derived from the inner mass of the mammalian blastocyst and have shown to be an exceptionally versatile source of stem cells with immense pluripotency and pro-liferative capacity (Thomson et al. 1998; Zulewski 2006). Soria et al. amongst the first teams successfully demonstrated that ESC are capable of maintaining a stable glucose response in streptozotocin (STZ)-induced diabetic mice in vivo (Soria et al. 2000). STZ is toxic to the insulin-producing β-cells in mammals, and therefore it is restricted to the engineering of animal model for T1D. The above study is analo-gous to recent studies by Clark et al. (2007) and Kroon et al. (2008), reporting the assessment of pancreatic endoderm derived from human embryonic germ (EG) cells and human ESC, respectively, with efficient production of glucose-responsive endocrine cells in vivo (in engrafted mice). Following transplantation into mice, differentiated cells have been identified by co-expression of transcription factors such as PDX1, FOXA2, SOX17, HNF6, and NKX6-1, marking a definitive pan-creatic endoderm. Approximately, 60–70 % of differentiated ESC produced the same level of glucose as compared to other studies in mice implanted with ∼3000–5000 adult human pancreatic islets (D'Amour et al. 2006). However, attempts to assess the amount of insulin generated by ESC have yielded disappointing results, with differentiated ESC under any specific culture condition producing only 1.6 % of the total insulin secreted by that of normal β-cells (Devaskar et al. 1994). Furthermore, despite the ESC capacity to differentiate into insulin-producing cells, there is a remote possibility that insulin-positive cells in ESC cultures may be a product of insulin uptake from the medium rather than an endogenous outcome. In addition, this approach is circumscribed, due to the ethical issues concerning the use of human ESC and their carcinogenic potential, low efficiency, and high cost.

ASC are further classified into hematopoietic stem cells (HSC) and mesenchy-mal stem cells (MSC), primarily obtained from the bone marrow (BM) with transdifferentiation plasticity, thus capturing great interest as a future therapy for T1D and other metabolic diseases. The promising role of MSC relies on two main aspects: first, they are multipotent and possess certain essential growth factors, which support tissue regeneration. Second, they present a broad range of immunomodulatory properties by interaction with a wide range of immune cells and hence this property can be harnessed to limit allograft rejection post-transplantation (Brusko 2009; Vija et al. 2009). Sun et al. demonstrated the in vitro differentiation of MSC into insulin-producing cells obtained from diabetic patients (Sun et al. 2007). The study further suggested that MSC could be used as a source of insulin, thereby reducing the risk of graft rejection. Voltarelli et al. conducted the first clinical trial in an attempt to explore the autologous potential of non-myeloablative HSC transplanted in T1D patients. A high-dose immunosuppression regimen was followed after post-transplantation (Voltarelli et al. 2007). After a mean follow-up period of approximately 9 months, increased C-peptide levels were seen in ~95 % of the patients, with adequate glycaemic control over a considerable period of time thereafter and with acceptable adverse effects reported.

As discussed earlier, MSC modulate the immunogenic properties of immune cells. In order to investigate this aspect, human-derived MSC (hMSC) were co-cultured with different subpopulations of immune cells in vitro, most notably T-cells. Here, hMSC influenced cytokine secretion levels with reduced levels of pro-inflammatory cytokines (TNF-α, IFN-γ, interleukin (IL)-6, and IL-17) and elevated levels of anti-inflammatory cytokines (IL-10, IL-4, and prostaglandin E2) (Aggarwal and Pittenger 2005; Vija et al. 2009; Ben-Ami et al. 2011). It has been corroborated that these mediators do indeed play a crucial role in the MSC-mediated immunosuppression. Another possible role of hMSC is to increase the T_{reg} cells level, which in turn down regulates T-lymphocyte proliferation and consequently suppresses autoimmunity (Aggarwal and Pittenger 2005; Augello et al. 2005). Furthermore, the question remains whether the effect exerted by MSC is due to the release of cytokines or due to cell-to-cell contact between MSC and T-cells (Werdelin et al. 1998; Aggarwal and Pittenger 2005; Augello et al. 2005; Selmani et al. 2008). There are several conflicting studies in the literature, which lead to the conclusion that there is most likely a range of mechanisms contributing to MSC-mediated immunomodulation.

At present, several human clinical trials deploying MSC-based therapies are underway. Osiris therapeutics is conducting a phase II study (Clinical trial no. NCT00690066) using ex vivo cultured ASC (Prochymal®) for treatment in recently identified, early onset T1D patients. Preclinical studies thus far have not shown any adverse effect of the allogenic MSC, and whether this remains the case in a clinical setting will become apparent as clinical trial data become available. To circumvent anticipated adverse events from allogenic MSC, much effort has been invested toward encapsulation techniques that protect the transplanted cells. Encaptra®, an advanced drug delivery system, is under development by Viacyte, Inc. Encaptra® consists of islet-like structures that release insulin in response to changes in physiological glucose levels in vivo. It is manufactured from FDA compliant implant grade materials, and is a SC implantation device designed with the goal to exclude the need for continuous immunosuppression.

Cell-based therapies are touching upon new undiscovered horizons and hence an increased understanding of the various factors contributing to β-cell mass dynamics is warranted. This will enable researchers to experiment with new approaches to induce β-cell mass expansion and promote their survival. The current knowledge of the field points toward seeking an absolute therapy for T1D, and the approaches outlined above, namely, islet formation, promotion of cell survival, and long-term maintenance of function, as well as the prevention of recurrent autoimmunity, should be examined simultaneously albeit judiciously.

4.4 Immune Intervention Strategies

From T1D pathophysiology, it is apparent that the primary physical indicator of β-cell destruction presents itself as insulitis, accompanied by the infiltration of $CD4^+$ and $CD8^+$ autoreactive cells. Based on murine studies, it is believed that during ongoing insulitis, ~ 80–90 % β-cells are damaged even before clinical diagnosis of T1D (Sreenan et al. 1999); this appears a dubious statement and only recently a meta-analysis study was carried out to gauge its validity Klinke (Klinke 2008). The conclusions drawn from the study paint a more detailed picture, indicating that the level of β-cell destruction varies with age and furthermore and depends on the insulin-producing capacity, which is proportional to β-cell mass, with body weight also playing a key role (Klinke 2008). Bearing these factors in mind, current immunotherapeutic approaches embody several innovative strategies, which include antibodies, antigens, and cytokines in individual- or cocktail-based therapies, and the section below provides an in-depth account of state-of-the-art approaches being pursued as treatment options for T1D.

4.4.1 Nonantigen-Based Intervention Strategies

Nonantigen-based intervention is primarily focused on the prevention of autoimmunity by modulating or silencing the immune response in the absence of negative effects on T_{regs} cells.

4.4.1.1 Immunosuppression

Immunosuppression is amongst the earliest reported intervention strategies used at the onset of T1D (see Table 4.1 for comprehensive list of immunosuppressive technologies); however, it has proven of limited use with only partial or in some cases no remission of T1D.

4.4.1.2 Antibody-Based Immunotherapy

Given that immunosuppressive intervention in T1D to date has failed to live up to its promise, it remains the challenge to find therapies that are not only highly efficacious but also safe when translated into humans, where exposure to exogenous

Table 4.1 Immunosuppressive agents in T1D and their mechanism of action

Immunosuppressive method/agents and MOA	References
Plasmapheresis • Partial or no success • Drawback: Need for immunosuppressive drugs, and a painful procedure hence not advisable in children	Ludvigsson et al. (1983), Hao et al. (1999)
Cyclosporine (CsA) • Maintains C-peptide and reduces HbA1c levels • Diabetes French study: 46 % remission rate in treatment group with few (minor) adverse effects reported • Drawbacks: Dose dependence, long-term malignancy, not optimal for children	Stiller et al. (1984), Feutren et al. (1986), Jenner et al. (1992), Schreiber and Crabtree (1992)
Nicotinamide (vitamin D) • High dose prevents or delays onset of T1D • DENIS and ENDIT studies failed to show any significant improvement in diabetes risk; No adverse effects observed	Gale et al. (2004), Kamal et al. (2006)

Cyclophilin CpN; *Calcineurin* CaN (serine/threonine phosphatase activity); cytoplasmic component of nuclear factor of activated T-cell; Nuclear component of nuclear factor of activated T-cell; *Poly* (ADP-ribose) *polymerase* PARP; *Nicotinamide adenine dinucleotide* NAD$^+$; *Poly* (ADP = ribose) PAR; *Deutsche Nicotinamide Intervention Study* DENIS; *European Nicotinamide Diabetes Intervention Trial* ENDIT antigens can be altogether avoided. The solution to this conundrum may well lie in the employment of a targeted approach, whereby monoclonal antibodies (mAb) can lead the way by directing T_{eff} cells, with the former also acting as biological immunomodulators (Li 2011). Some of the mAbs specifically designed for this purpose include anti-CD3, -CD20, -CD25, anti-thymocyte globulin (ATG), and CTLA-4, which target immune T_{eff} cells and many are presently under investigation in a range of T1D clinical trials.

T-Cell Targeting Antibodies

Anti-CD3 Antibody (Otelixizumab and Teplizumab)

Orthoclone-OKT3[®] (Muromonab-CD3 antibody) was the first US-FDA approved murine-derived anti-CD3 mAb, used in humans to prevent acute rejection post-organ transplantation. Preclinical studies in NOD mice showed that treatment with OKT3[®] resulted in potent reversal of T1D and successfully prevented immune responses toward the transplanted syngeneic islets (Herold et al. 1992; Chatenoud et al. 1994; Chatenoud and Bluestone 2007). As the name suggests, these mAbs are targeted to CD3 present on the surface of T-cells, which are necessary for T-cell activation. The immunomodulatory mechanism of CD3 mAb is suggested to act in two phases: phase I is attributed to T-cell depletion (preferentially T_{eff} cells) and activation of cells that secrete a range of cytokines (IL-10, TNF-α, IFN-γ). Phase II is proposed to have long-lasting effects, which involve induction of T_{reg} cells with enhancement of $CD25^{+}$ Forkhead box protein 3 (Foxp3) population, which is further dependent on transforming growth factor beta (TGF-β) -β) (Waldron-Lynch and Herold 2011). These postulated mechanisms render anti-CD3 a potential therapeutic agent; however, its side-effect profile includes transient cytokine syndrome which causes fever, hypotension, and arthralgia. The side effects were supposedly linked to activation of T-cells due to binding of murine F_c portion (fragment crystallizable region of antibody) of the antibody with the F_c receptors (F_cR) present on human cells (Abramowicz et al. 1989; Chatenoud et al. 1989).

In order to mitigate the effects of anti-CD3 mitogenic potential humanized, non-F_c binding anti-CD3 mAbs were engineered (Table 4.2). Teplizumab (humanized non-F_c-engineered mAb, also known as MGA031, hOKT3γ1, and Ala-Ala) and Otelixizumab (aglycosylated chimeric/humanized non-F_c binding mAb, also known as TRX4 and chAgCD3) have shown significant improvement in terms of safety due to their fully humanized profile, minimizing the potential for immunogenicity associated with the differences in mAb isotype between humans and mice (Keymeulen et al. 2005). Nonetheless, adverse events have been reported with anti-CD3, which include flu-like symptoms due to transient cytokine release as well as increased incidences of EBV infection following a single course of anti-CD3; this was, however, self-limiting with EBV incidence subsiding 1–3 weeks post-treatment initiation (Keymeulen et al. 2010a, b).

Table 4.2   Outcome of various anti-CD3 antibody-related clinical trials

Study title	Details	Results	Reference
Teplizumab mAb (Anti-CD3 mAb/hOKT3γ1/Ala–Ala)			
Autoimmunity-blocking antibody for tolerance in recently diagnosed type 1 diabetes (AbATE)	• Phase II; n = 81 (age range: 8–30 year) • Duration post-diagnosis: 8 weeks • Status: active • ClinicalTrials.gov Identifier: NCT00129259	Ongoing	Herold et al. (2009)
Protégé Study-clinical trial of MGA031 in children and adults with recent onset type 1 diabetes mellitus	• Phase II/III; n = 516 (age range: 8–35 year) • Duration post-diagnosis: 12 weeks • Status: complete • ClinicalTrials.gov Identifier: NCT00385697	β-cell preservation (detectable C-peptide level), lower insulin requirement	Sherry et al. (2011)
Anti-CD3 mAb treatment of recent onset type 1 diabetes (Delay)	• Phase II; n = 60 (age range: 8–30 year) • Duration post-diagnosis: 4–12 months • Status: complete • ClinicalTrials.gov Identifier: NCT00378508	Ongoing	http://clinicaltrials.gov/ct2/show/NCT00378508
Teplizumab for Prevention of Type 1 Diabetes In Relatives "At-Risk"	• Phase II; n = 140–170 (age range: 8–45 year) • Subjects with high risk • Status: recruiting • ClinicalTrials.gov Identifier: NCT01030861	Ongoing	http://clinicaltrials.gov/ct2/show/NCT01030861

(continued)

Table 4.2 (continued)

Study title	Details	Results	Reference
Otelixizumab mAb (Anti-CD3 mAb/TRX4/chAglyCD3)			
Extension of Phase II therapeutic trial with a humanized non-mitogenic CD3 (ChAgly CD3) monoclonal antibody in recently diagnosed Type 1 diabetic patients	• Phase II; n = 80 (age range: 12–45 year) • Duration post-diagnosis: 4–12 months • ClinicalTrials.gov Identifier: NCT00627146	Reduced insulin requirement over a period of 2 years, depending on their age and initial residual beta cell function	Keymeulen et al. (2010a, b)
Trial of Otelixizumab for adults with newly diagnosed type 1 diabetes mellitus (autoimmune): DEFEND-1	• Phase III; n = 272 (age range: 12–45 year) • Duration post-diagnosis: 90 days • Status: complete • ClinicalTrials.gov Identifier: NCT00678886	Failed to meet primary end point of change in C-peptide level at 12 months	http://www.gsk.com/media/pressreleases/2011/2011_pressrelease_10039.htm
Trial of Otelixizumab for adolescents and adults with Newly diagnosed type 1 diabetes mellitus (autoimmune): DEFEND-2	• Phase III; n = 396 (age range: 12–17 year) • Duration post-diagnosis: 90 days • Status: active but not recruiting • ClinicalTrials.gov Identifier: NCT01123083	Suspended pending reviews for DEFEND-1	http://clinicaltrials.gov/ct2/show/NCT01123083

Future developments of anti-CD3 antibodies should continue to focus on improving the safety along with increasing the efficacy. An additional factor to be considered is the stage of disease at the time of drug administration, as more promising results were observed in preclinical models when these agents were introduced in the earliest stages of the disease. Five independent large-scale clinical trials presently underway (phase II/III) have reported some encouraging outcomes (see the details in Table 4.2). The best response to therapy was noted in patients with a greater β-cell mass prior to initiating therapy; this implies that anti-CD3 therapy, if combined with agents that improve β-cell regeneration such as GLP-1 or other mimetics, may prove to be a superlative approach (Sherry 2007). One such attempt at this dual approach was reported with a combination of anti-CD3 antibody with intranasal proinsulin peptide, yielding encouraging results (Bresson et al. 2006).

Anti-CD25 (Daclizumab)

Another promising immunomodulatory antibody is anti-CD25 (Daclizumab), a humanized mAb which binds CD25, the α-subunit of the IL-2 receptor expressed on T-cells, and by doing so it inhibits the activation of T-cells. Daclizumab has been applied in autoimmune conditions such as multiple sclerosis (MS) and uveitis (Bielekova et al. 2009) although studies showing its effect in T1D monotherapy are absent. A recent study in diabetes-resistant bio-breeding (DR–BB) rat model demonstrated the synergistic effect of Daclizumab and Mycophenolate mofetil (MMF) (with MFF having a cytotoxic effect on autoreactive T-cells) (Ugrasbul et al. 2008). Disappointingly, a recent phase III clinical study using the same combination therapy failed to preserve β-cell function in new-onset T1D patients (Gottlieb et al. 2010). Moreover, although nonsignificant, a higher incidence of side effects was observed with the dual therapy, when compared to the placebo group.

Anti-thymocyte Globulin—Thymoglobulin®

Thymoglobulin® (anti-thymocyte globulin/ATG) is a purified γ-immune globulin raised in rabbits and horses via immunization against human thymocyte. This immunosuppressive agent contains cytotoxic antibodies directed against antigens expressed on human T-cells (Simon et al. 2008). Currently, rabbit-derived ATG is licensed as a treatment for acute rejection in renal transplant patients across Europe and USA. It is also indicated in therapies such as graft host disease (GVHD), aplastic anemia, and cancer. It was reported that in NOD mice, administration of murine ATG prevented progression of new-onset diabetes and induced long-term tolerance. The mechanisms involved are similar to anti-CD3, where first transient but substantial increases in cytokine release (TNF, IFN-α, IL-2, IL-3, and IL-6) (Ferran et al. 1991) are observed. Next, there is transient deletion of T-cells (T_{eff}) with increased frequency of $CD4^+CD25^+$ T_{reg} cells, although the levels of T_{regs}

cells were found to return to baseline after 30 days of treatment with ATG. In vitro and in vivo data support another plausible explanation, i.e., alteration of DC cells causing reduced expression of CD8$^+$ expression along with increased IL-10 production (K. Womer, unpublished study). Nonetheless, optimal outcomes with ATG therapy are dictated not only by dose, but also by the time from diagnosis and when therapy is initiated, with maximum efficacy seen when administration occurs after recent onset or late in the prediabetic phase (at 12 weeks of age). In complete contradiction to these findings, there is now evidence suggesting that the effect of ATG antibodies is dependent on the animal model employed (Bresson and Von Herrath 2011). This suggests that, before making any broad conclusion on the applicability of antiglobulin therapy, it is essential to assess it in various models of the disease, other than the most popular ones (e.g., NOD mice and BB rats) models.

Transitioning from trials in animals to those in humans, there are reports of a combination therapy, using ATG alongside prednisone. Positive results with lowered HbA1c levels and decreased insulin requirement per day were observed over several months. However, adverse effects of ATG therapy were also noted; these included cytokine release syndrome (leading to symptoms like fever and serum sickness) and even thrombocytopenia, which precluded its further use (Saudek et al. 2004). Currently, a trial entitled Study of Thymoglobulin to Arrest Newly Diagnosed Type 1 Diabetes (START) (Clinical trial ID: NCT00515099) aims to determine whether ATG treatment can halt the progression of newly diagnosed T1D when given within 12 weeks of disease diagnosis. Further, a combination trial in phase I/II has commenced using Thymoglobulin® and granulocyte colony-stimulating factors (GCSF) (Neulasta®) to identify their safety and potential to preserve insulin production. This combination therapy has shown durable remission of T1D in NOD mice due to the induction of T$_{reg}$ cells (Parker et al. 2009).

Cytotoxic T-Lymphocyte Antigen 4 Immunoglobulin (CTLA4-Ig)

The fusion immunoglobulin CTLA4-Ig (Abatacept®) acts on a number of costimulatory molecules (e.g., CD80 (aka B7-1) and CD86 (aka B7-2)) present on APCs (Fig. 4.2), functioning as a negative regulator of T-cell response (Perrin et al. 1995; Arima et al. 1996). T-cells require a secondary signal, in addition to the primary signal transmitted via MHC/TCR interaction, for their full activation. Hence, CTLA4-Ig acts by blocking CD28/B7 interaction (essential for the activation of T lymphocytes), leading to anergy (for further explanation, refer to the previous Sect. 3, Genetic susceptibility). Recently, blockade of the CD28/B7 costimulatory pathways has gained much attention, since it demonstrated to halt the progression of ongoing autoimmunity (Lenschow et al. 1992). An additional study reported that a single dose of CTLA4-Ig resulted in long-term graft survival after allogenic transplantation; this was marked by increases in cytokine IL-4 and IL-10, both of which are characteristic of Th2 subtype population. Separately, they were also found to induce production of TGF-β, a key regulatory cytokine involved in the immune regulation, thus supporting their role in the prevention of autoimmunity (Salomon et al. 2000; Salomon and Bluestone 2001; Tang et al. 2003). Furthermore, these immunoglobulins induce tolerance in allergenic islet transplantation by indolamine 2, 3-dioxygenase (IDO) mechanism (Fig. 4.3). IDO is a tryptophan degrading enzyme secreted by specific DCs, and hence it controls T-cell activation (Alexander et al. 2002). However, in certain cases, opposite to the CTLA-4 preventive action, it is found to augment autoimmune conditions. This could be due to (a) blocking of CTLA-4, which is a negative regulator of T-cell activation and (b) direct impact on the T_{regs} whose survival and differentiation depend on CD28 (Lenschow et al. 1996; Rigby et al. 2008). Hence, their blockade further results in exacerbation of T1D. The overall discussion so far seems to show that blocking CD28/B7 pathway on one hand may provide benefit in preventing or at least delaying the autoimmune disorder. On the other hand, its complex involvement in the maintenance of T_{regs} somehow makes its utility as an immune-modulating agent difficult. Other potential issues such as toxicity, time dependence, and site of administration need critical evaluation and to this end phase II clinical trials are underway using CTLA4-Ig (Abatacept®) in cases of new-onset T1D (Orban et al. 2011) and LEA29Y (Belatacept®) in islet transplantation (Chatenoud 2011).

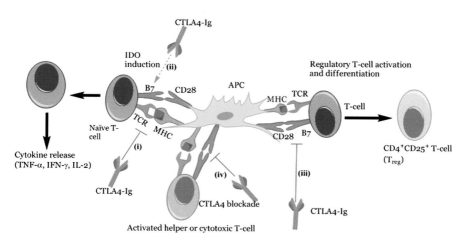

Fig. 4.3 Different roles of CTLA4-Ig implicated in immune response. CTLA4-Ig exerts its action via several different mechanisms: (**i**) CTLA4-Ig blocks the CD28 interaction with B7 owing to its high affinity for B7 co-receptor and results in blocking of T-cell activation or anergy, (**ii**) Induction of IDO following interaction of CTLA4-Ig with B7 which prevents T-cell activation, (**iii**) Immune suppression effect due to the blockade of CD28 resulting in reduction of Th2 cell differentiation, i.e., T_{reg} cells and finally, (**iv**) Inhibits the CTLA-4 binding with B7 ligands necessary for CTL activation which are essential to control T_{eff} cell function

B-Cell Targeting Antibody

Literature abounds with illustration of the clear effector role of T-cells in autoimmune T1D and hence most of the therapeutic studies target T-cells for treatment purposes. From our previous discussion, we know that NOD mice demonstrate a plausible role of B-cells as APC in T1D pathogenesis, more than just producing antibodies against autoantigens. This somehow indicates that therapies such as anti-CD20 and anti-CD25, which target B-cells, may have a positive role in the disease remission. In fact, anti-CD20 mAb treatment in NOD mice has shown depletion of CD20 B-cells, which prevented T1D in more than 60 % of tested mice population due to their predominant interference with $CD4^+$ T-cell activation (Serreze and Silveira 2003).

Anti-CD20 (Rituximab)

Rituximab is a chimeric mAb targeted against CD20 transmembrane receptors, found primarily on the surface of B-cells (Looney 2005; Martin and Chan 2006). Rituximab was first introduced in 1997 and is the first therapeutic mAb, approved by the European Medical Agency (EMEA) and US-FDA, for the treatment of B-cell non-Hodgkin's lymphoma (NHL). It is additionally used in other therapies such as rheumatoid arthritis, systemic lupus erythematosus (SLE), and multiple sclerosis (MS) (Pescovitz et al. 2011). A key advantage of rituximab is its specificity for CD20 molecules compared to other treatment strategies, which prevents undesired side effects. A clinical trial sponsored by TrialNet included 87 patients (age range: 8–40 years) with newly diagnosed T1D, who received 4 weekly infusions of rituximab. The study reported that, after 1 year, stimulated C-peptide level was found to be 20 % higher in the rituximab-treated group compared to the placebo group. Secondary treatment responses included lowered HbA1c level and lowered daily doses of insulin with no increase in infection or neutropenia observed. Unfortunately, the C-peptide response was short lived with maximal drug effect being apparent only in the early months of treatment (Pescovitz et al. 2009). This suggests that rituximab may not be an optimal approach for T1D treatment but certainly it provides a proof-of-concept that B-cell depleting agents are a perfectly feasible target for halting the ongoing progression in T1D. However, more meticulous investigation is a prerequisite to identify the benefit-to-risk ratio, which seems more delicate in the case of children suffering from T1D. Additionally, the translation from preclinical to clinical studies has shown substantial variations, further rendering this approach a far-fetched idea.

4.4.1.3 Cytokine-Based Therapy

Cytokines are small, cell signaling molecules which orchestrate different processes such as T-cell growth (interleukins IL-2, IL-4, IL-7, IL-15, and IL-21), inflammatory responses by Th1/Type 1 cytokines (IL-1, TNF-α, TNF-β, IFN-α, and IFN-γ), and anti-inflammatory function by Th2/Type 2 cytokines (IL-4, IL-10, and TGF-β), respectively. Further, Th2 cytokines are involved in the expansion of T_{reg} cells. The binary role of cytokines as mediators and regulators opens new prospects for T1D therapy. Some paradoxes exist the assumed inflammatory role of Th1 cytokines, since the administrations of IL-2 and IFN-γ are actually found to prevent T1D in animal models (Rabinovitch and Suarez-Pinzon 2007).

Interleukin 1 (IL-1)

IL-1, a pro-inflammatory cytokine, is a significant and overexpressed cytokine gene in peripheral blood mononuclear cells (PBMC) in patients with newly diagnosed T1D compared to healthy controls (Kaizer et al. 2007). In NOD mice, systemic knockout of IL-1 receptor gene was found to reduce diabetes mellitus rates by 30 % (Thomas et al. 2004). Anti-IL-1 (Anakinra) is a recombinant, non-glycosylated human version of anti-IL-1 receptor antagonist, clinically approved for the treatment of rheumatoid arthritis. In 2007, a clinical trial showed that blockade of IL-1 pathway through the use of Anakinra improved glycemia, β-cell secretory function, and reduced markers of systemic inflammation in T2D (Larsen et al. 2007). Recently, a study related to use of anti-IL-1 therapy in children with newly diagnosed T1D was conducted. Preliminary results showed that anti-IL-1 therapy is well tolerated in children. A similar level of HbA1c and mixed-meal tolerance test was observed in the treated and control groups, with additional low dose of insulin requirement in treatment group, 1 and 4 months post-diagnosis (Sumpter et al. 2011). The limitations of this study include non-randomized experiments, limited sample size, brief duration of therapy (28 days), independent controls groups (selected differently), and no precedent study to support the therapy duration. Nevertheless, the study suggests the need for randomized, placebo-controlled study over a longer duration in order to evaluate the true capacity of anti-IL-1 therapy to augment the insulin secretory capacity of T1D patients. Another study including anti-IL-1 mAb (Canakinumab) is currently in progress (ClinicalTrials.gov ID: NCT00947427).

Interleukin 2 (IL-2)

IL-2 is one of the cytokines exhibiting several functions such as T-cell growth, cytotoxic effects, and maintenance of T_{reg} cells. Malfunctioning of IL-2 signaling in $CD4^+$ T-cells of T1D subjects leads to lowered Foxp3 expression on the surface of T_{reg} cells, lowering their tolerance effect (d'Hennezel et al. 2010). Bearing this theory in mind, IL-2 was tested preclinically and was found to be dose dependent. It was observed in NOD mice that low-dose administration of IL-2 resulted in increased induction of T_{reg} cells with higher expression of T_{reg}-related proteins such as Foxp3, CD25, CTLA-4, ICOS (inducible T-cell costimulator), and GITR (glucocorticoid-induced TNF receptor), and prevented T1D. Furthermore, IL-2 administration suppressed IFN-γ production (Grinberg-Bleyer et al. 2010). A cytolytic chimeric IL-2 and F_c fusion protein with specific binding to IL-2 receptor was found to suppress induced T1D in NOD mice (Zheng et al. 1999), while the F_c portion of the protein contributed to lowered immunogenicity and increased half-life ($t_{1/2}$). According to Clinical Trials.gov, a study was recently completed using IL-2 (Aldesleukin) (study ID: NCT01353833) aimed at assessing its dose–effect relationship on T1D with results yet to be declared (http://clinicaltrials.gov/ct2/show/NCT01353833). A further study using a combination of IL-2 and rapamycin has shown limited success (Long et al. 2012). In contrast to the limited success with cytokines treatment, blockade of pro-inflammatory cytokines seems a more pertinent approach (Baumann et al. 2012).

Tumor Necrosis Factor-Alpha (TNF-α)

In addition to the aforementioned cytokines, TNF-α is additional pro-inflammatory cytokine which exhibits direct cytotoxic effects on β-cells. Anti-TNF therapy has gained wide recognition in the treatment of rheumatoid arthritis and Crohn's disease, in particular. However, only recently it has been tested in various clinical trials evaluating their efficacy in T1D. In NOD mice, TNF-α or anti-TNF-α therapies have shown contrasting results in vitro and in vivo, with ablation of T1D in vivo as opposed to exacerbation of T1D or destruction of β-cells in vitro. This suggests that TNF-α might have a more potent immunoregulatory effects in vivo than a mediator effect (Jacob et al. 1990). In addition, the role of TNF-α in T1D pathogenesis remains enigmatic. One of the initial studies showed the age-dependent effect of TNF-α therapy in NOD mice. Administration of TNF-α (before 4 weeks of age) led to an earlier onset of T1D, whereas total blockade of TNF-α with mAb prevented T1D. The outcome clearly demonstrates the indispensable role of TNF-α in the thymic development of autoimmune repertoire, which could be due to its involvement in thymocyte activation and proliferation (Yang et al. 1994; Christen and Herrath 2002; Kodama et al. 2005; Koulmanda et al. 2012). In an attempt to study the feasibility and efficacy of anti-TNF-α therapy, a soluble recombinant TNF fusion protein (Etanercept) was used in a pilot scale study on 18 children with newly diagnosed T1D. The preliminary data showed prolonged effects on endogenous insulin production in these pediatric patients (Mastrandrea et al. 2009). Further, large-scale studies are underway to evaluate the true potential and associated side-effect profile with the use of this therapy.

Striking differences noticed with the use of cytokines reveal the need for careful contemplation based on (i) time of administration, (ii) route of administration, (iii) dose, and (iv) site of expression (local, i.e., organ-specific or systemic) (Bresson and Von Herrath 2007). A strategy such as localization of cytokine therapy directly to the site of islet inflammation is also something which needs attention, so as to avoid nonspecific and undesirable side effects in vivo. In summary, these data mark the complexity of using cytokine alone as immune intervention. Finally, there is an enormous potential that the anti-cytokine treatment, along with other therapies, may prevent or delay diabetes development in individuals at high risk of the disease.

4.4.2 Antigen-Based Intervention Strategies

Antigen-based therapy is a developing field and currently relishes the most attention. The rationale following its investigation is antigen specificity which adds to its safety profile, something not found with immunosuppressive agents or nonantigen-specific immunomodulation approaches (Ludvigsson 2009). However, these therapies suffer drawbacks such as unpredictable outcome of immunization that depends on numerous factors such as antigen dose, frequency, and route of administration (Peakman and von Herrath 2010).

Treatment using autoantigen delivery is a well-known concept and dates back to the twentieth century when it was used to treat hypersensitive allergies toward innocuous foreign antigens (Krishna and Huissoon 2011). The specific induction of T_{regs} in response to administration of an exogenous antigen has been termed inverse or negative vaccination (Mercer and Unutmaz 2009; Hinke 2011). By poorly understood mechanisms, naive T-cell activation matures either into T_{eff} or $CD4^+CD25^+$ T-cell/T_{reg} cells. As we know from our previous discussion, T_{eff} cells are responsible for induction of an immune response. T_{regs} can be further classified as (i) natural, i.e, thymus derived (nT_{reg}) or, (ii) adaptive, i.e, induced within the periphery (iT_{reg}). These T_{regs} are essential for the maintenance of immune tolerance. In some cases, the Foxp3, a marker of T_{reg} cells, shows increased expression on these cells. Unfortunately, these Foxp3 are murine-specific T_{reg} markers, because their utility in humans is restricted as they are additionally expressed on recently activated T_{eff} cells. The fate of T-cells depends on antigenic signal strength and prevalent cytokines. The underlying mechanism behind this therapy is the tendency of autoantigens to either delete the effector/pathogenic (Th1 and CTLs) and autoreactive ($CD4^+$ and $CD8^+$) T-cells or their ability to expand protective T_{reg} specific to the autoantigens (Bresson and von Herrath 2009). Several T_{reg}-associated mechanisms have been implicated in the development of immune tolerance by antigen-specific therapy. First, bystander suppression which refers to specific antigen induced generation of T_{reg} cells that nonspecifically alter the function of APC and hence suppress inflammation in the target organs by suppressing the T_{eff} cells induced by other antigens (Anderton et al. 1999; Millington et al. 2004). Second, infectious tolerance which means T_{reg} cell clone has a natural ability to induce T_{regs} of different antigen specificities (Li 2011; Harrison 2012).

Several ways are available to achieve antigen-specific therapy of T1D in animal models: (i) use of full length islet proteins (such as GAD65), (ii) altered peptide ligands derived from islet antigens (such as heat shock protein 6), and/or (iii) with DNA vaccines (InsB$_{9-23}$) (Slobodan et al. 2011). Various autoantigens associated with T1D like hsp60/Diapep277, insulin (Ins), and GAD65 (Diamyd®) have been tested in clinical trials with varied levels of success. The next section includes details of the updated knowledge of different antigen-specific therapies.

4.4.2.1 Insulin Vaccine

Insulin and proinsulin epitopes have been identified as playing a vital role in the early stages of T1D autoimmune processes both in humans and NOD mice models (Zhang et al. 1991; Kent et al. 2005; Nakayama et al. 2005). Insulin autoantibodies (IAA) precede the clinical presentation of T1D especially in children and are often used as biomarkers in disease prediction. Insulin therapy has shown a twofold benefit: (i) it restores the insulin-specific immune tolerance and, (ii) it halts ongoing β-cell destruction with the supply of active, exogenous insulin, and thus (iii) it reduces the insulin secreting stress on the pancreas. Based on the large body of evidence which shows insulin's role in T1D, inverse vaccination with insulin and proinsulin either as a peptide or DNA vaccine seems like an ideal and relatively facile approach for prevention or intervention in T1D (Hilsted et al. 1995; Skyler 2008). Table 4.3 includes various trials that are complete or in ongoing stages using insulin as a drug candidate.

4.4.2.2 Heat Shock Protein 60(Hsp-60)/Diapep277

DiaPep277 is a 24 amino acid-containing peptide derived from positions 437-460 of major T-cell epitope of Hsp60. Hsp60 functions in protein folding as an intracellular chaperone and it is a vital stimulator of the innate immune system. Hsp has been found to activate the APCs like macrophages and DCs by activating the pro-inflammatory response via Toll-like receptor 4 (TLR4), while it is also believed to induce an anti-inflammatory response via TLR2 (Anderton et al. 1993) (Fig. 4.4). In contrast to the inflammatory action of Hsp, DiaPep277 only activates the anti-inflammatory response TLR2 and hence shifts the inflammatory Th1 response to protective Th2 immune response (Tuccinardi et al. 2011). DiaPep277 has been modified to be stable in vivo by replacement of two cysteines at locations 6 and 11 of the sequence, respectively, by valine. The replacement does not induce any physiological change in the native DiaPep277 sequence (Raz et al. 2001). Inflammation inhibitory activity of Hsp60 was first reported in mycobacteria-immunized rats. Ragno et al. in late 1990s demonstrated the protective effect of Hsp60 as a naked DNA vaccine in adjuvant arthritis (Ragno et al. 1997; Quintana et al. 2003). Soon afterward, mammalian Hsp60 was found as a

Table 4.3 Outline of prevention and intervention trials using insulin/proinsulin as a target antigen

Study/clinical trial	Therapy stage	Antigen	Dose/follow-up period/route of administration	Study group/age group	Study outcome	Lesson learnt	References
Pre-POINT	Primary prevention	Oral/intranasal insulin	• Insulin daily for 1st 10 days followed by twice dose every week • 2.5–67.5 mg (Oral) • 0.28–7.5 mg (Intranasal) Mean follow-up of ~3–18 months	1st degree T1D relatives with >50 % risk of T1D Age: 1.5–7 years	Recruiting	• Dose escalating study Intend to find optimal intervention timing and dose	Rewers and Gottlieb (2009)
DPT-1 (1st arm)	Secondary prevention	Oral insulin	• Dose: 7.5 mg/day • Mean follow-up of ~4.3 years	1st and 2nd degree T1D relatives with 25–50 % risk of T1D Age: <5 years	• Same C-peptide and HbA1c level in treatment and control group • No reduction in insulin requirement	• Lack of insulin efficacy • Selection of patient with variable T1D risk (39–79 nU/ml IAA level) • Insulin degradation administered orally • Insufficient dose	Skyler et al. (2005)
DPT-1 (2nd arm)	Secondary prevention	Crystallized recombinant insulin (ultralente)	• Dose: total dose of 0.25u/kg/day BID (SC) + annual 4 days (IV) insulin infusion • Mean follow-up of ~3.7 year	1st and 2nd degree T1D relatives with 50 % risk of T1D Age: <5 year	• Similar cumulative incidence of T1D in treatment and control group (RR = 0.96 in treatment group) • No delay or prevention of T1D in high risk group	• Intervention timing	(Diabetes Prevention Trial–Type 1 Diabetes Study 2002)
TrialNet (subgroup analysis of DPT-1 with high IAA level >80nU/ml)	Secondary prevention	Oral insulin	• Dose: 7.5 mg/day	1st and 2nd degree T1D relatives	• Delayed T1D progression for ~4.5 years. • No reduction in IAA levels	• Arbitrary and low dose may not affect the immune system • Different method of autoantibody assay	Skyler et al. (2005), Barker et al. (2007)

(continued)

Table 4.3 (continued)

Study/clinical trial	Therapy stage	Antigen	Dose/follow-up period/route of administration	Study group/age group	Study outcome	Lesson learnt	References
DIPP	Secondary prevention	Intranasal insulin	• Dose: 1U/kg body weight per day • Mean follow-up of ~2 years	Children (<3 year of age) with HLA-conferred susceptibility to T1D	No prevention or delay of T1D	• Insufficient dose • Need more safe and efficacious antigen-specific therapies	Näntö-Salonen et al. (2008)
INIT I	Secondary prevention	Intranasal insulin	• Dose: twice 400 μg of insulin per day for 10 days • Mean follow-up of ~3 years	At-risk T1D relatives (median age of ~ 10.8 years)	Well tolerated and stable β-cell function	• Identification of safe intranasal insulin dose	Harrison et al. (2004)
INIT II	Secondary prevention	Intranasal insulin	• Dose: two doses of 4 IU and 440 IU per day for 7 days + weekly for a year • Mean follow-up of ~4 year	T1D relatives aged 4–30 years with autoantibodies to at least 2 islet antigens (~40 % risk of diabetes over 5 years)	Recruiting	–	http://www.nhmrc.gov.au/_files_nhmrc/media_releases/2011 0205/166_06.pdf; Harrison (2008)
Diabète insuline orale group	Tertiary prevention/Intervention	Recombinant insulin	• Oral + intense SC insulin therapy	Recent onset of T1D	No prevention in β-cell residue	• Insufficient dose • Formulation choice leading to early insulin degradation (lower bioavailability) • End stage intervention • Low study sample	Chaillous et al. (2000)
IMDIAB VII	Tertiary prevention/Intervention	Oral insulin	• Dose: 5 mg/day (oral) + intense SC insulin therapy • Mean follow-up of ~1 year	Recent onset of T1D	Similar insulin requirement post 1 year with similar HbA1c and IAA levels in the treatment and control group	• No effect may be due to low dose • Low β-cell mass at time of T1D diagnosis	Pozzilli et al. (2000)

(continued)

Table 4.3 (continued)

Study/clinical trial	Therapy stage	Antigen	Dose/follow-up period/route of administration	Study group/age group	Study outcome	Lesson learnt	References
NBI-6024 (phase I/II)	Tertiary prevention/Intervention	APL Ins$_{9-23}$	*Phase I:* • 5 biweekly (0.1, 1 or 5 mg) SC • Mean follow-up of 6 months *Phase II:* • 26 doses (1st 3 doses administered biweekly followed by monthly doses) SC • Mean follow-up of 24 months	After new onset of T1D *Phase I:* Age: adolescent (13–16 years) Adults (20–41 years) *Phase II:* Age: adolescent (10–17 years) Adults (18–35 years)	*Phase I:* Shift of Th1 pathogenic response to Th2 protective regulatory phenotype (significant ↑ in IL-5 response). Well tolerated *Phase II:* No improvement or maintenance of β-cell function. Reported ≥10 % patient with severe adverse events but not suspected to be drug related	• *Phase I:* Safety and tolerance in Phase I lead to multidose study • *Phase II:* Lack of response may be due to fundamental defect in proposed MOA or due to dose, frequency, or timing of intervention	Alleva et al. (2006), Walter et al. (2009)
IBC-VS01	Tertiary prevention/Intervention	Insulin B-chain peptide + IFA	*Phase I* • 2 mg InsB-chain peptide in IFA introduced once IM • Mean follow-up of 2 years	In new onset of T1D (n = 12) Age: 18–35 years	• Safe and well tolerated. • Long-lasting, robust immune response with generation of autoantigen-specific T_{reg} cell capable of arresting T1D	• Pilot study hence lack of power. • Warrants further studies	Orban et al. (2010)
Proinsulin peptide	Tertiary prevention/Intervention	C19-A3	• 30 or 300 μg in 3 monthly doses ID • Mean follow-up of 6 months	Patients with long-standing T1D Age: 21–53 years	• Well tolerated and safe. • Increased IL10 with induction of proinsulin-specific T_{reg} cell	Warrants further studies in new-onset T1D patients	Thrower et al. (2009)

(continued)

Table 4.3 (continued)

Study/clinical trial	Therapy stage	Antigen	Dose/follow-up period/route of administration	Study group/age group	Study outcome	Lesson learnt	References
BHT-3021	Tertiary prevention/Intervention	DNA plasmid encoding proinsulin	• 1 of 4 dose levels (0.3, 1, 3 or 6 mg) weekly for 12 weeks IM • Mean follow-up of 25–37 months	After new onset of T1D Age: ≥18 years	Interim results: For 1 mg dose, 50 % ↓ in anti-insulin Ab titer in treated patient. After 15 weeks no significant clinical abnormality and loss of β-cell function	No adverse event	http:// clinicaltrials. gov/ct2/show/ NCT0045375

Diabetes Prevention Trial of Type 1 Diabetes DPT-1; bis in die BID; Subcutaneous SC; Relative risk RR; Intravenous IV; International Diabetes Immunotherapy Group IMDIAB; Insulin autoantibody IAA; Type 1 Diabetes Prediction and Prevention Study DIPP; Intranasal insulin trial INIT; POINT Primary Oral/Intranasal Insulin Trial; Antigen peptide ligand APL; Incomplete Freund's adjuvant IFA; Month mo; Year Yr; Intradermal ID; Intramuscular IM; Mechanism of action MOA
Primary prevention targets the young children carrying high genetic HLA-DR, DQ genotypic risk of T1D; Secondary prevention targets the population who have autoimmunity with no disease but carry risk of development of T1D; Tertiary prevention or intervention done when T1D is diagnosed but there is a window of opportunity to prevent the C-peptide and β-cell loss

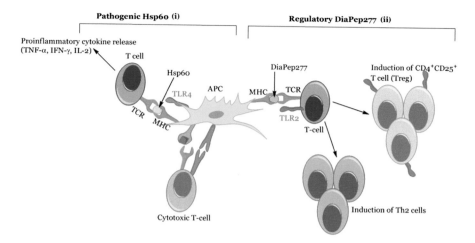

Fig. 4.4   Role of Hsp60 (**i**) as a pathogenic autoantigen and (**ii**) mechanism of action of DiaPep277 vaccine

target of T-cells involved in the etiology of T1D (Elias et al. 1990). Antibodies to this Hsp60 autoantigen have been detected in new-onset T1D patients. However, it is difficult to unravel their role as a predictive marker of T1D (Abulafia-Lapid et al. 1999). Apparently, DiaPep277 is involved not only in immunomodulation of the innate immune system, but also in adaptive immune responses. This discovery has given rise to its application as an autoantigenic vaccine, which is expected to help maintaining its specificity. Preservation of residual β-cells in NOD mice has been observed even in the progressive or late stages of autoimmunity (Elias and Cohen 1995). In addition to the clinical trials mentioned in Table 4.6, two different trials were run on latent autoimmune diabetes of adults (LADA) patients with inconclusive results. Table 4.6 includes all available clinical studies, from phase I to III.

4.4.2.3   Glutamic Acid Decarboxylase 65 (GAD65)

GAD is a 585 amino acid-containing enzyme encoded at chromosome 10p11, responsible for the synthesis of a vital inhibitory neurotransmitter in the central nervous system, namely γ-amino butyric acid (GABA). Disturbed GAD production is responsible for lowered GABAergic signaling, leading to seizures and stiff person syndrome (SPS). High anti-GAD antibody titers are found in SPS and T1D patients (Levy et al. 1999). However, it should be noted that the epitopes of GAD identified in both the diseases show different isotypic patterns (Lohmann et al. 2000). GAD is correspondingly located in the pancreas, though its specific role within pancreas remains enigmatic. Some studies suggest possible involvement of GABA in the hormone secretion within the pancreas, while some indicate its involvement in the hormone secretion in response to glucose. Principally, autoantibodies to two iso-forms of GAD have been identified in T1D, namely GAD65 and GAD67 (with molecular weight 65 and 67 kDa, respectively). GAD65 was identified as autoantigen for the first time in 1982 using plasmapheresis, with the immunopre-cipitation of a protein with a molecular weight of 64 kDa (Baekkeskov et al. 1982, 1990; Karlsen et al. 1991). Subsequent studies confirmed the presence of autoan-tibodies to this antigen in T1D and several autoimmune disorders. Nearly, 60–70 % of the T1D populations are found to possess autoantibodies to GAD65 antigen, and this is especially the case in adult patients (Kaufman et al. 1992; Han et al. 2011). The presence of GAD65 has been identified as one of the important biomarkers of T1D. Preclinical studies in murine models have indicated that mucosal, intraperi-toneal (IP), IM, and SC administration of formulated GAD65 prevents the devel-opment of autoimmunity (Gong et al. 2010). Functionally, immunization with GAD65 induces a shift of Th1 response to Th2 type along with induction of subset of T_{reg} cells, which eventually prevents or reduces long-term disease incidence (Tian et al. 1996).

More than 20 preclinical studies were performed to evaluate the safety and efficacy of GAD65 and no undesirable effects were observed indicating a wide safety margin associated with its use. GAD65-Alum (Diamyd®) was used as a human adjuvant-based vaccine due to clinical reasons, which include the following: (i) alum has been used commercially in vaccines for more than 70 years, (ii) alu-minum salts are preferential humoral immune response inducers rather than cellular response inducers, and (iii) alum is the only adjuvant contained in vaccines licensed by the US-FDA for human use (Sesardic et al. 2007; Dekker et al. 2008; Uibo and Lernmark 2008). Details of the various clinical studies are given in the table below (Table 4.4). Failure in phase III trials with GAD65 alum is without a doubt an unfortunate setback, but nevertheless these studies sufficiently indicate the persis-tent need to have better understanding of the dose, route of administration, and most essentially the stage of intervention in order to maximize the induction of immune tolerance. However, several subsequent trials such as a Swedish prevention trial in children with high risk of T1D and interventions using a cocktail of agents with GAD65 are underway.

4.4.3   Cell-Based Therapy

In addition to the broad range of approaches discussed in the earlier sections, one of the most vibrant fields witnessing its horizon in T1D immunotherapy approach is cell-based therapy. The rationale behind this therapy is to compensate for the lost pancreatic β-cells through (i) replacement of islet cells by islet-like cells derived from different stem cells (like ESC, HSC, or ASC) and (ii) restoration of immunological tolerance to islet pathogenic self-antigens through adoptive transfer

Table 4.4 Summary of clinical trials using antigen-specific therapy

Immunotherapy	Clinical Trial/Entry criteria	Study outcome	Lesson learnt	References
Hsp60$_{437-460}$ (DiaPep277)	Phase II; newly diagnosed (<6 months) T1D (n = 35;16–58 years) Dose: 10 months treatment with 1 mg DiaPep277 + 40 mg mannitol in vegetable oil administered SC at 1 and 6 months	C-peptide production maintained. Lower insulin requirement	• Small scale study • Safe and well tolerated • Study was continued for longer (found similar outcomes)	Elias and Cohen (1995), Raz et al. (2001, 2007)
Hsp60$_{437-460}$ (DiaPep277) (two arm trial)	Phase II; Recent onset of T1D (<3 months) *Adult group*: (n = 50; 16–44 years) Dose: 0.2, 1 or 2.5 mg at 0,1,6 and 12 months SC injection *Pediatric group*: (n = 49; 4–15 years) Dose: 0.2 and 1 mg at 0,1,6 and 12 months SC	*Adult group*: At 18 months modest maintenance of β-cell function in 0.2 and 1 mg group; significant loss of C-peptide in 2.5 mg *Pediatric group*: At 13 months stable C-peptide in 1 mg group; at 18 months pronounced β-cell loss than adults	• Age is determining factor for β-cell and depends on HLA-susceptibility and level of immunity which is weak in children	Schloot et al. (2007), Huurman et al. (2008)
Hsp60$_{437-460}$ (DiaPep277)	Phase II; Children with recent onset of T1D (< 53-116 days); n = 30; 7–14 years Dose: 1 mg at 0,1,6 and 12mo SC	At 18mo, similar C-peptide level in both group. No metabolic control	• Small sample size • May need increased frequency of antigen dose	Lazar et al. (2007)
DIA-AID (Andromeda 901)	Phase III; Newly diagnosed T1D (< 3 months) (n = 457; 16–45 years); C-peptide ≥0.2 nmol/L Dose: 1 mg + 40 mg mannitol in 0.5 ml lipid emulsion SC injection at 0,1,3,6,9,12,15,18 and 21 months. *Clinicaltrials.gov ID: NCT00615264*	• Significant preservation of C-peptide level compared to placebo group • Significant higher HbA1c level maintained in treated group	• Safe and well tolerated • Additional studies to determine the long-term effect and maintenance	Raz et al. (2014)

(continued)

Table 4.4 (continued)

Immunotherapy	Clinical Trial/Entry criteria	Study outcome	Lesson learnt	References
DIA-AID (extension study)	Phase III: Patient completing study 901 with additional DiaPep277 treatment (n = 50; 18–48 years); C-peptide ≥0.2 nmol/L Additional follow-up of 2 years Clinicaltrials.gov ID: NCT00644501	Recruiting Aim: To evaluate safety with extended treatment and long-term efficacy	–	http:// clinicaltrials. gov/ct2/
DIA-AID II	Phase III; Newly diagnosed T1D (<6 months) (20–45 years) Dose: 1 mg SC injection at 0,1,3,6,9,12,15,18,21 and 24 months Clinicaltrials.gov ID: NCT01103284	Recruiting Aim: To study the treatment effect of DiaPep277 on preservation of β-cell function, as defined by meal-stimulated secretion of insulin	–	http:// clinicaltrials. gov/ct2/
rhGAD65 (Diamyd UK)	Phase I; Healthy volunteers (n = 16) Dose: 20–500 μg single SC injection	No significant treatment related adverse effects. No induction of autoantibodies towards GAD65, IAA and IA-2	• Safe and well tolerated • Promoted escalating dose studies	http://www. diamyd.com
rhGAD65-Alum (Diamyd®)	Phase II: LADA (n = 47) Dose: 4,20,100 and 500 μg SC at week 1 and 4 Mean follow-up of 24 weeks	↑ fasting C-peptide and ↑ CD4+CD25+/CD4+CD25− T-cell subset ratio in 20 μg treated group but not in other groups	• Clinically safe • 20 μg as effective dose	Kaufman et al. (1993), Tisch et al. (1999), Casas et al. (2007)

(continued)

Table 4.4 (continued)

Immunotherapy	Clinical Trial/Entry criteria	Study outcome	Lesson learnt	References
rhGAD65-Alum (Diamyd®)	Phase II: Children and adolescent (n = 70; 10–18 years) C-peptide ≥0.1 nmol/L + detectable GAD antibodies. Dose: 20 μg SC at week 1 and 4. Mean follow-up of 30 months. Clinicaltrials.gov ID: NCT00435981	• Maintenance of fasting and stimulated C-peptide over 30 months • A 4-year follow-up showed better preserved C-peptide in treated patients with < 6 months T1D	Safe and well tolerated. No treatment related adverse event	Ludvigsson et al. (2008, 2011)
rhGAD65-Alum (Diamyd®)	Phase II: Recently diagnosed T1D (<100 days) (n = 126; 3–45 years). Dose: 20 μg SC at week 1,4, and 8 (3 doses) or 20 μg SC at wk 1 and 4 (2 doses). Clinicaltrials.gov ID: NCT00529399	Similar loss of insulin secretion at 1 year between treated and placebo group	• Vaccine may prove beneficial if used in earlier disease stages • Can prove effective in combination therapy	Wherrett et al. (2011)
rhGAD65-Alum (Diamyd®)	Phase III; Newly diagnosed T1D (<3 months) (n = 334; 10–20 years); C-peptide >0.1 nmol/L + detectable GAD autoantibodies. Dose: 20 μg GAD-Alum SC on day 1,30,90 and 270 (4 doses),or 20 μg GAD-Alum SC on day 1 and 30 + 20 μg placebo on day 90 and 270 (2 dose + 2dose) or 4 doses of 20 μg of placebo on day 1,30, 90 and 270. Clinicaltrials.gov ID: NCT00723411	• No prevention of C-peptide loss in any treatment group at 15 months. • No effect on insulin dose, HbA1c or hypoglycemia rate. Mild adverse events	• Population difference and clinicians with different conventional treatment approach • Speculate seasonal variation • Age and sex difference as analyzed in subgroup exploratory analysis	Ludvigsson et al. (2012)

Streptozotocin STZ; Latent autoimmune diabetes in adult LADA; Intraperitoneal IP of $CD4^+CD25^+$ T_{reg} and targeting APCs such as DC cells. Since islet regeneration and stem cell therapy have been discussed in detail earlier, the following section is focused on establishment of immunological tolerance methods.

4.4.3.1 $CD4^+CD25^+Foxp3$ T-Cell

Propelled by the large body of evidence obtained from studies in animal models, $CD4^+$ T_{regs} quintessentially appears to play a protective role in homeostasis maintenance, which otherwise progresses into autoimmune conditions such as T1D (Kabelitz et al. 2008; Spoerl and Li 2011). Growing data suggest the likely existence of several subsets of $CD4^+$ T_{reg} cells, amongst which the most investigated ones fall in two classes: (1) $CD4^+CD25^+Foxp3$ T_{reg} and (2) Th2 cells secreting anti-inflammatory cytokines (IL-10 and TGF-β) (Grazia Roncarolo et al. 2006). $CD4^+CD25^+Foxp3$ T_{reg} constitutes 5–10 % of total peripheral $CD4^+$ T-cells, with high expression of CD25 and Foxp3 which plays a pivotal role in maintaining homeostasis and immune tolerance (Bluestone and Tang 2004; Fontenot et al. 2005; Zheng and Rudensky 2007). Development of this cell lineage is believed to be from two different origins, i.e, natural T_{reg} (nT_{reg}) (which are developed within the thymus) and another known as adaptive or induced T_{reg} (iT_{reg}), generated in the periphery (in vitro or in vivo) in response to antigenic stimuli in specific milieu.

A myriad of molecular and cellular mechanisms have been implicated in T_{reg}-mediated immune suppression. The translational effect of these T_{reg} is believed to be due to their direct or indirect interaction with APCs resulting in either bystander suppression or infectious tolerance (Qin et al. 1993; Thornton et al. 2004). These two essential mechanisms allow the T_{regs} with limited-antigen specificity to establish a broad and stable immunoregulatory effect and the required T_{reg} phenotype. The other mechanisms of note include (1) weak stimulation via T-cell and APC interaction, which leads to suppressed pro-inflammatory (such as IL-2) cytokine secretion; (2) release of soluble factors such as protective cytokines (IL-10 and TFG-β); (3) cell-to-cell contact-dependent mechanism whereby suppression is induced by T-cell accessory molecules such as CTLA-4 (direct suppression or via IDO expression) (Vignali et al. 2008) and lymphocyte activation gene (LAG3), expressed on T_{regs}, while CD80 and CD86 expressed on the cell surface of APC (Sakaguchi 2004; Miyara and Sakaguchi 2007), and (4) cytolytic effect of naturally occurring T_{reg} cells expressing granzyme A and perforin, which leads to apoptosis of T_{eff} cells (Grossman et al. 2004a, b; Gondek et al. 2005). Several other mechanisms have also been shown to orchestrate T_{reg} suppression; however, the precise mechanism remains elusive.

A crucial problem faced in T_{reg} research is identifying a characteristic cell-surface marker for human T_{regs} that will aid their easy identification and isolation from the set of $CD4^+$ T-cells. Largely, mice and human T_{regs} were isolated based on the expression of CD25, found to be expressed on the activated T-cells and not primarily on T_{regs} alone (Yi et al. 2006). Similarly, Foxp3 has been identified as an indispensable regulator involved in T_{reg} differentiation, development, function, and their identification in mice (Fontenot et al. 2005); however, they are not human-specific markers. Mutation in the Foxp3 gene results in malfunction and lowered number of nT_{regs}, followed by autoimmune diseases such as scurfy disease in mice (Lahl et al. 2007) and IPEX syndrome in humans (Gambineri et al. 2003). Furthermore, recent studies have shown that a low expression of CD127, together with high expression of CD25, is a reliable marker for delineating Foxp3$^+$ nT_{regs} (Liu et al. 2006). This suggests that the search for specific T_{reg} marker has remained elusive, requiring further extensive research.

The discussion so far leads to the fact that the alteration or dysfunction of T_{reg} cell is associated with various autoimmune diseases, including T1D (Thomas et al. 2005). Adoptive transfer of CD4$^+$CD25$^+$ T_{reg} cells in NOD mice has shown to prevent autoimmune T1D (Bluestone and Tang 2004; Masteller et al. 2005; Tarbell et al. 2007) development and several other metabolic diseases such as SLE, irritable bowel syndrome (IBD), MS, etc. Following preclinical studies, the first-in-human clinical effects of adoptive transfer of ex vivo expanded CD4$^+$CD25$^+$CD127$^-$ T_{regs} was found to alleviate disease symptoms in patients suffering from acute or chronic GVHD (Cao et al. 2009; Trzonkowski et al. 2009). Current efforts have been directed toward inducing and maintaining tolerance using therapeutic vaccination with CD4$^+$CD25$^+$ T_{reg} cells, which can be achieved either directly or indirectly (through the use of anti-CD3 antibody or antigen-specific immunotherapy) (Bluestone and Tang 2004). A recent pilot study conducted in newly diagnosed T1D children showed for the first time that the infusion of ex vivo expanded autologous T_{regs} prolongs disease remission (Marek-Trzonkowska et al. 2012). Several approaches for in vivo induction of suppressive adaptive T_{reg} have also been investigated. One such example is humanized anti-CD3 antibody administration that has shown beneficial effects on residual pancreatic β-cell function (Belghith et al. 2003). Several other approaches using antigen alone (GAD65 and insulin B_{9-23}) (Mukherjee et al. 2003) or simultaneously with anti-CD3 antibody are also under investigation (Bresson et al. 2006). Various research labs are focused on developing optimal methods to expand the antigen-specific T_{regs} in vitro and many have been successful in doing so. For example, Tang et al. have been able to develop a robust method to expand the T_{regs} in vitro by 200-fold within 2 weeks using a combination of anti-CD3, anti-CD28, and IL-2 (Tang et al. 2004), while a small number of associated T_{regs} were also found to reverse diabetes suggesting their utility in T1D immunotherapy.

Nevertheless, the successful translation of T_{reg} therapy in a clinical setting is thwarted by several challenges, including isolation and purification of T_{regs} from peripheral blood of patients, expansion technologies in vitro and in vivo, and determination of their long-term stability in the adoptively transferred host and safety issues in particular, related to the polyclonal T_{regs}. Antigen-specific immunotherapy somehow seems a plausible resolution to the problem of specificity with still much room for improvement.

4.4.3.2   Dendritic Cells

Earlier discussion briefly suggests that antigen-specific therapy holds the potential as key player in the coming future for T1D therapeutics. Henceforth, a marked interest has evolved toward the manipulation of DCs that are identified as the most potent APC in T1D and other autoimmune conditions (Steinman 2008). DCs are involved in different stages of T1D progression, as they are the first cells to infiltrate the pancreas followed by their role in the activation of autoreactive T-cells, respectively. Conversely, in steady state, DCs are found to have a low MHC expression and costimulation molecules that allow them to present an antigen, however, without immunogenic response induction (Steinman et al. 2003). This indicates that in steady and/or immature state, DCs maintain tolerance S induction of anergy or T_{reg} cells.

Current efforts involving DCs as therapeutics aim at exogenous generation of DCs for administration as a vaccine. One of the clinical trials has recently shown the safety aspect of the exogenous DC-based vaccine for application in T1D (Clinical trial.gov identifier: NCT00445913) (Giannoukakis et al. 2011). This study was planned based on preclinical data, which demonstrated that BM-derived DCs treated ex vivo with a mix of antisense oligonucleotides selectively knocked down the costimulatory molecules, especially CD40, CD80, and CD86. The costimulatory impairment was able to prevent and reverse T1D in NOD mice model (Machen et al. 2004). An alternative attractive approach involves the in vivo targeting of DCs with polymeric microparticles that possess the capacity to deliver the vaccine to DCs (Zhao and Leong 1996; Keselowsky et al. 2011). One of the most comprehensively studied molecules for targeting Ag to DC is DEC-205. DEC-205 includes endocytic receptors, expressed at high levels on lymphoid tissue DCs and is found to significantly enhance the efficiency of antigen presentation. The following approach involved the ligation of DEC-205 targeted antigen with that of mAb, which acts as stimuli for maturing DCs. It was shown that such a targeted approach led to higher and effective T-cell-mediated immunity (Bonifaz et al. 2004)

However, so far this DC-based vaccine only provides a proof-of-concept, as several factors limit this approach, such as complex cellular isolation and storage with high costs involved. Nonetheless, there are still unanswered questions that remain before the clinical application of this technology such as route of DCs administration and frequency of injections (Morelli and Thomson 2007).

4.5   Combination Therapies in Type 1 Diabetes, the Way Forward?

Delving in literature reveals that almost a dozen of clinical trials were completed with little or no success (Table 4.5). In light of the current evidence, it may not be altogether in appropriate to suggest that current monotherapies may fall short of current expectations as far as T1D therapies are concerned. The involvement of multiple heterogeneous factors significantly complicates disease control and prevention; hence, it is imperative to target multiple biological pathways involved in the disease. It is very well understood that ideal intervention for T1D would include not only entities that can halt the autoimmune response, but perhaps an additional element that will help to enhance β-cell function and/or expedite β-cell regeneration. Based on current preclinical and clinical data, future successes firmly appear to lie with combination therapies. These combination/cocktail therapies offer several advantages over the current therapies, which include improved safety, dose reduction, and synergistic effects that further prolong efficacy. Recently, the *Juvenile Diabetes Research Foundation* (JDRF) and *Immune Tolerance Network* (ITN) have made a concerted effort toward developing strategies that will prioritize the transition to human trials based primarily on combination therapies (Boettler and von Herrath 2010; Matthews et al. 2010). However, in order to proceed further, both sufficient and reliable safety and efficacy data on the use of various monotherapies first need to be made available. Separately, there is an additional challenge of engaging large pharma industries, who carry the main cost burden, in combination drug trials, especially where the individual drugs are already marketed, or are in the developmental stages so disclosure and IP protection issues then emerge as matters of concern. Nevertheless, several promising opportunities exist for the use of drug combinations using approved drugs. For example, anti-inflammatory drugs (IL-1; anti-TNF-α), T-cell modulators (anti-CD3), B-cell depleting agents (anti-CD20), antigen-specific agents (oral insulin; GAD-alum; proinsulin DNA or peptide), and incretin mimetic (exenatide) along with islet transplantation are preferred choices in cocktail therapies (Bresson and Von Herrath 2007). However, none of the combination therapies tested so far have shown success in the clinical phase as can be seen in Table 4.5.

4.6   Lesson Learnt from Animal Models

It is now well accepted that animal models of disease are essential to better understand the global effects of an agent or combination of agents, species differences aside. To this end, the accessibility to reliable experimental models for the investigation of T1D pathogenesis with additional requirements for determination of safety and efficacy profiling is readily available for T1D (Brett et al. 2011). From a practical perspective, animal models are relatively easier to work with, not least due to issues of accessibility in humans to the pancreas and islets, given their location deep in the peritoneum. So the bulk of our existing knowledge pertaining to T1D pathology stems from rodent models, which develop spontaneous T1D, the most notable being NOD mice and BB rats due to their features closely resembling

Table 4.5 Combination therapy clinical trials: completed, ongoing, and planned

Combination therapy	Mechanism	Study in animal models	Clinical trials	Results and commentaries	References
Exenatide + Daclizumab (CD25) + intense insulin therapy	Blockade of IL-2 signaling pathway and β-cell growth	No data on combination in preclinical setting	Phase II: (n = 20) patients tested with long-standing T1D (21.3 ± 10.7 years) *Clinicaltrials.gov ID: NCT00064714*	No improvement in C-peptide levels	Rother et al. (2009)
Daclizumab (CD25) (SC) + Mycophenolate mofetil (MMF) (oral)	Blockade effect of CD25 inhibits IL-2 pathway and cytotoxic effects of MMF on autoreactive T-cells	Synergistic effect of combination therapy found in DR–BB rats	Phase III: (n = 126) patients chosen within 3 months of diagnosis and with sufficient C-peptide *Clinicaltrials.gov ID: NCT00100178*	• Geometric mean of C-peptide AUC over 2 years • No effect of drugs alone or in combination	Ugrasbul et al. (2008), Gottlieb et al. (2010)
IL-2 + Rapamycin (sirolimus) (officially titled: Proleukin (SC) + rapamune (oral)	Shift of Th1 cells to Th2/Th3 type cells due to selective inhibition of IL-2 signaling	Proven efficacious in NOD mice	Phase I: (n = 9) patients recruited within 3– 48 months of diagnosis; *Clinicaltrials.gov ID: NCT00525889*	• Transient β-cell dysfunction despite an increase in T_{regs} • Need careful translation with intense investigation on immune system that effects therapy	Rabinovitch et al. (2002), Long et al. (2012)

(continued)

Table 4.5 (continued)

Combination therapy	Mechanism	Study in animal models	Clinical trials	Results and commentaries	References
Diamyd (GAD65) + Lanzoprazole + Sitagliptin (DPP inhibitor)	GAD specific immunomodulation and proton pump inhibitor provide partial reversal of diabetes	Drugs individually proven efficacious	Phase II; double blind, parallel study, safety and efficacy *Clinicaltrials.gov ID: NCT00837759*	Study terminated	Agardh et al. (2009), Kim et al. (2009)
Anti-CD3 + Nasal or oral insulin	Immunomodulation with islet antigen-specific induction of T_{regs}	Synergistic effect seen in treating recent onset T1D in NOD and RIP–LCMV mice	Planned	N/A	Bresson et al. (2006)
ATG (thymoglobulin®) + GM-CSF (Neulasta®)	GM-CSF induces tolerogenic DCs and both ATG and GM-CSF induces T_{regs}	Reversal at new-onset diabetes with combination therapy in NOD mice (blood glucose level up to 500 mg/dl)	Phase I/II; *Clinicaltrials.gov ID: NCT01106157*	Recruiting patients	Parker et al. (2009)
EGF (E1) + Gastrin (G1)	Increases β-cell regeneration and neogenesis	Normoglycemia achieved with administration in early onset diabetes (Blood glucose level > 200 mg/dl)	Phase I; (n = 20) patients; *Clinicaltrials.gov ID: NCT00239148*	Completed; result processed recently (N/A)	http://clinicaltrials.gov/ct2/show/NCT00837759

(continued)

Table 4.5 (continued)

Combination therapy	Mechanism	Study in animal models	Clinical trials	Results and commentaries	References
Anti-CD3 + IL-1 receptor antagonist	Induction of T_{reg} by anti-CD3 mAb and IL-1 receptor antagonist negatively affects the differentiation of CD4$^+$ T-cells	Synergistic effect in NOD mice. Improved insulin level and reduced number of T_{eff} cells	–	–	Ablamunits et al. (2012)
CTLA4-Ig + InsB$_{9-23}$	Blockade of costimulation required for T-cell activation along with T_{reg} cell induction	Failure to reverse the new-onset of T1D	–	Showed the significance of testing the combination drugs in animal models before translation in humans	Schneider et al. (2012)

Clinicaltrials.gov ID CTG ID; *IL-2* Interleukin -2; *Th1* T helper cell 1; *Th2* T helper cell 2; *Th3* T helper cell 3; *DR–BB* Diabetes-resistant bio-breeding; *DPP* Dipeptidyl peptidase; *ATG* Anti-thymocyte globulin; *GM-CSF* Granulocyte macrophage colony-stimulating factor; *EGF* Epidermal growth factor; *CTLA4-Ig* Cytotoxic lymphocyte antigen 4 immunoglobulin; *InsB$_{9-23}$* Insulin chain B (residue 9–23)

Table 4.6 Enumeration of the similarities and differences between NOD mice and humans

Properties	NOD mice	Human
Presence of Mφ, DC, CD4$^+$, CD8$^+$ and B-cells	Yes	Yes
Genetic susceptibility loci	Orthologue of I-Ag7 allele	MHC class II alleles
Insulitis phenotype	Extensive and easily spotted	Few leukocyte detected
Autoantibody	Insulin	GAD, IA-2 and insulin
Incidence and prevalence of T1D (male versus female)	Higher in females	Similar
Life span	2 years	∼78 years
Life pattern	Inbred strain from genetically identical animals under specific conditions	Heterogeneous population

that in humans (see Table 4.6). Additionally, other murine models such as knockout models, humanized murine models, and transgenic models have also been tried and tested (Bresson and von Herrath 2009). Amongst all murine models, NOD mouse is an autoimmune strain which has been extensively exploited to dissect the lack of immune tolerance and autoimmunity in T1D (Fornari et al. 2011). Inter-individual heterogeneity in humans complicates the pathogenesis of T1D and challenges the determination of true prevalence rates; this is further complicated by the dearth of efficient surrogate markers and imaging techniques. Often results obtained from NOD mice have been overtly extrapolated into humans, e.g., the fiasco resulting from anti-CD28 therapy (Suntharalingam et al. 2006). In spite of the discrepancy risk, preclinical animal models have proven invaluable, offering crucial information toward better understanding the etiopathogenesis of T1D.

To date out of 230 intervention strategies assessed in murine models, less than 10 (or ≈4 %) have demonstrated any kind of clinical effect in human trials. This exceptionally poor rate of success suggests that investigators need to seriously consider the underlying causes for the immunological and genetic disparities, and consider a broader range of doses and frequency of dosing, the environment in which experiments are conducted and whether their interpretation may be further substantiated through the use of parallel investigations in other animal models (e.g., transgenic or larger animals) (Roep and Atkinson 2004). In order to encompass the aforementioned criteria, in future trials, and strengthen any association to human subjects, it is likely that researchers and clinicians will need to broaden their collaborations, in order to meet the increased demands of the additional parameters in expanded studies.

4.7   Perspective of Nanotechnology in Type 1 Diabetes Drug Delivery

Nanotechnology is not a fully developed concept but has gained a lot of attention in the last few decades in various domains of science and technology. Application of nanotechnology to medicine is used at a scale length of $\approx 1–100$ nm range, with the creation and use of structures, formulations, devices, and systems that have novel properties and functions owing to their miniaturized size (Gordon et al. 2003). Pharmaceutical companies have been focusing nowadays to develop targeted drug delivery using nanotechnology, with the specific aim to reach the target cells. Although this area of science is still at its horizon, it already presents an incredible prospective for nanomedicine.

Following our previous discussion on the treatment of T1D, it is evident that, except for the daily administration of insulin and vaccination with anti-CD3 Abs, no other treatment has shown clinical success so far. With this view in mind, the quest to eliminate needles for insulin delivery with some alternative and noninvasive routes has driven many researchers toward new alternatives. One of the promising approaches in this direction is the development of transdermal drug delivery. Transdermal drug delivery is an appealing alternative to subcutaneous insulin delivery due to enhanced patient compliance and controlled release over time, while avoiding drug degradation in the GIT or first-pass liver effects. Albeit the offered advantages, there is a big hurdle for the successful drug delivery of insulin through skin, i.e., the presence of a horny layer known as *stratum corneum*. However, in the current scenario there are several approaches available to optimize the delivery, including chemical penetration enhancers, as well as electrical (iontophoresis) or physical (sonophoresis) methods. Another new concept added to this field is the development of microneedles that have shown effective delivery of protein and peptide drugs with reduced pain and damage to cells in skin (Pickup et al. 2008). However, translation of this system to be used in humans is still in progress.

Furthermore, nanoparticle carriers have also been tried and tested with limited success: examples include encapsulation of islets in the case of islet transplantation and coating of insulin for oral delivery. In particular, encapsulation of insulin with polymers such as poly-lactic acid (PLA), poly(lactic-co-glycolic acids) (PLGA), poly(alkylcyanoacrylate), and polymethacrylic acid nanoparticles has shown higher drug encapsulation efficiency leading to improved glycemia in diabetic rats (Damgé et al. 2008). In one such study, Sai et al. (1996) showed that feeding 100 IU/kg insulin nanocapsules to NOD mice reduced the incidence of diabetes and the severity of lymphocytic inflammation of endogenous islets. However, few nanoparticulate systems containing insulin have been investigated in humans. Generex Bioctechnology Corp. has tested its buccal insulin mouth spray (Oralin™). The formulation controlled postprandial glucose levels in both T1D and T2D patients. Others such as Emisphere's SNAC-insulin capsule formulation and

Nobex's hexyl-PEG-modified peroral insulin (Cernea et al. 2005) were tested in phase I and II clinical trials.

Besides the quest for treatment of T1D, another hot area of research in this field is diabetes management. It is essential to manage and monitor the individual blood glucose level to abate the ghastly negative effects of T1D. Conventional glucose measuring instruments suffer drawbacks of painful sampling and huge fluctuations due to missed sampling (Cash and Clark 2010). As an answer to this, nanotechnology and nanomedicine have been incorporated into the field of diabetes sensing and management. The sensors could be designed at macro-, micro-, and nanoscale, respectively. However, nanosensors would give an edge over macrosensors since they could be implanted. Furthermore, because of their nanosize, they might avoid immune response by the immune system and hence can have a long-term benefit. Such progress in the field of nanotechnology certainly implicates a bright future for the treatment of T1D. Techniques such as encapsulation of islets will drive the transplantation procedure offering a longer duration treatment.

4.8 Conclusion and Perspectives

T1D is primarily an autoimmune attack on the pancreas by autoreactive T-cells that results in comprehensive destruction of insulin-producing β-cells. Increasing incidence of T1D compels for better treatment options. The search for treatment option is complicated due to the additional factors contributing to the disease pathology such as individual genetics and environmental factors. Furthermore, an incomplete understanding of involved mechanism adds to the problem. Nevertheless, the primary treatment for T1D still involves daily administration of insulin. Alternative approaches to the intravenous administration of insulin include buccal, pulmonary, nasal, and transdermal routes. Second, interesting results have been obtained with other methodologies such as islet and pancreas transplantation, which increases or replaces the destroyed pancreatic β-cells.

Third, a shift has now been witnessed via replacement of conventional methods with immunotherapy. One of the successful approaches includes the administration of targeted anti-CD3 Abs that demonstrated improved C-peptide levels and lowered requirements of insulin. It is worth to notice that, albeit insignificant beneficial effects observed with individual immunotherapeutic agents, none has proven to be a breakthrough. Hence, it seems more worthwhile to investigate combination of drugs such as anti-CD3 and intranasal proinsulin peptide, which demonstrated to be more efficacious against monotherapies either with anti-CD3 alone or antigen alone in two different diabetes models. As described in the previous sections, the majority of the immunomodulatory drugs have failed to meet their primary end point in some phases of clinical trials. The probable reasons suggested for disappointments can be summarized as follows: (i) the administration route and dosage were not optimal;

(ii) variation in regulatory immune cell responses to peptides/Abs than those found physiologically; (iii) different timings of administration and stage of disease (based on residual β-cell mass) when the immunotherapeutic agents were administered; and (iv) lack of preclinical data such as in the case of combination therapies.

Henceforth, it is clear that T1D will be a pandemic if not attempted to be curbed in a strategic manner (Narayan et al. 2006). Having said this, it is of utmost importance to have a better understanding of the pathophysiology of the disease. Additionally, there is still a lot of room for improvement, which involves investigation of the appropriate target molecules and disease-associated factors. Another aspect of the disease that needs thorough investigation is to identify the stage of disease, age of patients, and duration of treatment in order to optimize the beneficial effects of any therapy. In other words, it is essential for the scientific community to have a widespread and efficient collaboration between research institutes and pharma industries. Unless the lessons learnt from laboratory level studies are translated into preclinical studies, it is difficult to draw conclusions on the efficiency and safety of the drug molecules. Furthermore, use of nanotechnology in the field of T1D is still in its incipient stage, despite some encouraging progress. Overall, nanotechnology is perceived to become a key tool in troubleshooting many of the T1D-related problems.

References

Ablamunits V, Henegariu O et al (2012) Synergistic reversal of type 1 diabetes in NOD mice with anti-CD2 and interleukin-1 blockade: evidence of improved immune regulation. Diabetes 61 (1):145–154

Abramowicz D, Schandene L et al (1989) Release of tumor necrosis factor, interleukin-2, and gamma-interferon in serum after injection of OKT3 monoclonal antibody in kidney transplant recipients. Transplantation 47(4):606–608

Abulafia-Lapid R, Elias D et al (1999) T cell proliferative responses of type 1 diabetes patients and healthy individuals to human hsp60 and its peptides. J Autoimmun 12(2):121–129

Agardh CD, Lynch K et al (2009) GAD65 vaccination: 5 years of follow-up in a randomised dose-escalating study in adult-onset autoimmune diabetes. Diabetologia 52(7):1363–1368

Aggarwal S, Pittenger MF (2005) Human mesenchymal stem cells modulate allogeneic immune cell responses. Blood 105(4):1815–1822

Alexander AM, Crawford M et al (2002) Indoleamine 2,3-dioxygenase expression in transplanted NOD Islets prolongs graft survival after adoptive transfer of diabetogenic splenocytes. Diabetes 51(2):356–365

Alleva DG, Maki RA et al (2006) Immunomodulation in type 1 diabetes by NBI-6024, an altered peptide ligand of the insulin B epitope. Scand J Immunol 63(1):59–69

Anderton S, Burkhart C et al (1999) Mechanisms of central and peripehral T-cell tolerance: lessons from experimental models of multiple sclerosis. Immunolo Rev 169(1):123–137

Anderton SM, van der Zee R et al (1993) Inflammation activates self hsp60-specific T cells. Eur J Immunol 23(1):33–38

Arima T, Rehman A et al (1996) Inhibition by CTLA4Ig of experimental allergic encephalomyelitis. J Immunol 156(12):4916–4924

Augello A, Tasso R et al (2005) Bone marrow mesenchymal progenitor cells inhibit lymphocyte proliferation by activation of the programmed death 1 pathway. Eur J Immunol 35(5):1482–1490

Baekkeskov S, Aanstoot HJ et al (1990) Identification of the 64 K autoantigen in insulin-dependent diabetes as the GABA-synthesizing enzyme glutamic acid decarboxylase. Nature 347(6289):151–156

Baekkeskov S, Nielsen JH et al (1982) Autoantibodies in newly diagnosed diabetic children immunoprecipitate human pancreatic islet cell proteins. Nature 298(5870):167–169

Banting FG, Best CH (1922) The internal secretion of the pancreas. J Lab Clin Med 7:251–266

Barker J, McFann K et al (2007) Effect of oral insulin on insulin autoantibody levels in the diabetes prevention trial type 1 oral insulin study. Diabetologia 50(8):1603–1606

Baumann B, Salem HH et al (2012) Anti-inflammatory therapy in type 1 diabetes. Curr Diab Rep 12(5):499–509

Beattie GM, Otonkoski T et al (1997) Functional beta-cell mass after transplantation of human fetal pancreatic cells: diiferentiation or proliferation? Diabetes 46(2):244–248

Belghith M, Bluestone JA et al (2003) TGF-beta dependent mechansims mediate restoration of self-tolerance induced by antibodies CD3 in over autoimmune diabetes. Nat Med 9(9):1202–1208

Ben-Ami E, Berrih-Aknin S et al (2011) Mesenchymal stem cells as an immunomodulatory therapeutic strategy for Autoimmune Diseases. Autoimmun Rev 10(7):410–415

Bielekova B, Howard T et al (2009) Effect of anti-CD25 antibody daclizumab in the inhibition of inflammation and stabilization of disease progression in multiple sclerosis. Arch Neurol 66 (4):483–489

Bluestone JA, Tang Q (2004) Therapeutic vaccination using CD4 + CD25 + antigen-specific regulatory T cells. Proc Natl Acad Sci USA 101(Suppl 2):14622–14626

Boettler T, von Herrath M (2010) Immunotherapy of type 1 diabetes - How to rationally prioritize combination therapies in T1D. Int Immunopharmacol 10:1491–1495

Bonifaz LC, Bonnyay DP et al (2004) In vivo targeting of antigens to maturing dendritic cells via the DEC-205 receptor improves T cell vaccination. J Exp Med 199(6):815–824

Brand SJ, Tagerud S et al (2002) Pharmacological treatment of chronic diabetes by stimulating pancreatic beta-cell regeneration with systemic co-administration of EGF and gastrin. Pharmacol Toxicol 91(6):414–420

Brusko TM (2009) Mesenchymal stem cells: a potential border patrol for transplanted islets? Diabetes 58(8):1728–1729

Bresson D, Togher L et al (2006) Anti-CD3 and nasal proinsulin combination therapy enhances remission from recent-onset autoimmune diabetes by inducing Tregs. J Clin Invest 116 (5):1371–1381

Bresson D, von Herrath M (2007) Moving towards efficient therapies in type 1 diabetes: to combine or not to combine? Autoimmun Rev 6(5):315–322

Bresson D, von Herrath M (2009) Immunotherapy for the prevention and treatment of type 1 diabetes. Diabetes Care 32(10):1753–1768

Bresson D, von Herrath M (2011) Anti-thymoglobulin (ATG) treatment does not reverse type 1 diabetes in the acute virally induced rat insulin promoter-lymphocytic choriomeningitis virus (RIP-LCMV) model. Clin Exp Immunol 163(3):375–380

Breton M, Farret A et al (2012) Fully integrated artificial pancreas in type 1 diabetes: modular closed-loop glucose control maintains near normoglycemia. Diabetes 61:2230–2237

Brett P, Massimo T et al (2011) Current state of type 1 diabetes immunotherapy: incremental advances, huge leaps, or more of the same? Clin Dev Immunol 2011:432016

Buckingham B, Chase HP et al (2010) Prevention of nocturnal hypoglycemia using predictive alarm algorithm and insulin pump suspension. Diab Care 33:1013–1017

Cao T, Soto A et al (2009) Ex vivo expanded human CD4 + CD25 + Foxp3 + regulatory T cells prevent lethal xenogenic graft versus host disease (GVHD). Cell Immunol 258(1):65–71

Casas R, Hedman M et al (2007) Diabetes 67th annual scientific sessions. 1242

Cash KJ, Clark HA (2010) Nanosensors and nanomaterials for monitoring glucose in diabetes. Trends Mol Med 16(12):584–593

Cernea S, Kidron M et al (2005) Dose-response relationship of an oral insulin spray in six patients with type 1 diabetes: a single-center, randomized, single-blind, 5-way crossover study. Clin Ther 27(10):1562–1570

Chaillous L, Lefèvre H et al (2000) Oral insulin administration and residual ([beta]-cell function in recent-onset type 1 diabetes: a multicentre randomised controlled trial. Lancet 356(9229): 545–549

Chatenoud L (2011) Diabetes: type 1 diabetes mellitus–a door opening to a real therapy? Nat Rev Endocrinol 7(10):564–566

Chatenoud L, Bluestone JA (2007) CD3-specific antibodies: a portal to the treatment of autoimmunity. Nat Rev Immunol 7(8):622–632

Chatenoud L, Ferran C et al (1989) Systemic reaction to the anti–t-cell monoclonal antibody OKT3 in relation to serum levels of tumor necrosis factor and interferon-α. N Eng J Med 320 (21):1420–1421

Chatenoud L, Thervet E et al (1994) Anti-CD3 antibody induces long-term remission of overt autoimmunity in nonobese diabetic mice. Proc Natl Acad Sci USA 91(1):123–127

Christen U, Herrath MG (2002) Apoptosis of autoreactive CD8 lymphocytes as a potential mechanism for the abrogation of type 1 diabetes by islet-specific TNF-α expression at a time when the autoimmune process is already ongoing. Ann NY Acad Sci 958(1):166–169

Clark GO, Yochem RL et al (2007) Glucose responsive insulin production from human embryonic germ (EG) cell derivatives. Biochem Biophys Res Commun 356(3):587–593

Damgé C, Reis CP et al (2008) Nanoparticle strategies for the oral delivery of insulin. Expert Opin Drug Deliv 5:45–68

D'Amour KA, Bang AG et al (2006) Production of pancreatic hormone–expressing endocrine cells from human embryonic stem cells. Nat Biotechnol 24(11):1392–1401

Dekker CL, Gordon L et al (2008) Dose optimization strategies for vaccines: the role of adjuvants and new technologies, NVAC subcommittee on vaccine development and supply, Washington, DC, HHS: 1–21

Devaskar SU, Giddings SJ et al (1994) Insulin gene expression and insulin synthesis in mammalian neuronal cells. J Biol Chem 269(11):8445–8454

Dor Y, Brown J et al (2004) Adult pancreatic beta-cells are formed by self-duplication rather than stem-cell differentiation. Nature 429(6987):41-46

d'Hennezel E, Kornete M et al (2010) IL-2 as a therapeutic target for the restoration of Foxp3 + regulatory T cell function in organ-specific autoimmunity: implications in pathophysiology and translation to human disease. J Trans Med 8(1):113

Elias D, Markovits D et al (1990) Induction and therapy of autoimmune diabetes in the non-obese diabetic (NOD/Lt) mouse by a 65-kDa heat shock protein. Proc Natl Acad Sci USA 87 (4):1576–1580

Elias D, Cohen IR (1995) Treatment of autoimmune diabetes and insulitis in NOD mice with heat shock protein 60 peptide p277. Diabetes 44(9):1132–1138

Elleri D, Dunger D et al (2011) Closed-loop insulin delivery for treatment of type 1 diabetes. BMC Med 9:120

Ellis S, Naik RG et al (2008) Use of continuous glucose monitoring in patients with type 1 diabetes. Curr Diab Rev 4:207–217

Engström HA, Johansson R et al (2008) Evaluation of a glucose sensing antibody using weak affinity chrmoatography. Biomed Chromatogr 22:272–277

Ferran C, Dy M et al (1991) Inter-mouse strain differences in the in vivo anti-CD3 induced cytokine release. Clin Exp Immunol 86(3):537–543

Feutren G, Papoz L et al (1986) Cyclosporin increases the rate and length of remissions in insulin-dependent diabetes of recent onset. Results of a multicentre double-blind trial. Lancet 2 (8499):119–124

Fontenot JD, Rasmussen JP et al (2005) Regulatory T cell lineage specification by the forkhead transcription factor foxp3. Immunity 22(3):329–341

Fornari TA, Donate PB et al (2011) Development of type 1 diabetes mellitus in nonobese diabetic mice follows changes in thymocyte and peripheral t lymphocyte transcriptional activity. Clin Dev Immunol 2011:158735

Fujinami RS, von Herrath MG et al (2006) Molecular mimicry, bystander activation, or viral persistence: infections and autoimmune disease. Clin Microbiol Rev 19(1):80–94

Gale EA, Bingley PJ et al (2004) European nicotinamide diabetes intervention trial (ENDIT): a randomised controlled trial of intervention before the onset of type 1 diabetes. Lancet 363 (9413):925–931

Gambineri E, Torgerson TR et al (2003) Immune dysregulation, polyendocrinopathy, enteropathy, and X-linked inheritance (IPEX), a syndrome of systemic autoimmunity caused by mutations of FOXP3, a critical regulator of T-cell homeostasis. Curr Opin Rheumatol 15(4):430–435

Giannoukakis N, Phillips B et al (2011) Phase I (safety) study of autologous tolerogenic dendritic cells in type 1 diabetic patients. Diab Care 34(9):2026–2032

Gondek DC, Lu LF et al (2005) Cutting edge: contact-mediated suppression by CD4 + CD25 + regulatory cells involves a granzyme B-dependent, perforin-independent mechanism. J Immunol 174(4):1783–1786

Gong Z, Pan L et al (2010) Glutamic acid decarboxylase epitope protects against autoimmune diabetes through activation of Th2 immune response and induction of possible regulatory mechanism. Vaccine 28(24):4052–4058

Gordon N, Sagman U et al (2003) Nanomedicine taxonomy. CIHR IRSC:1–32

Gottlieb PA, Quinlan S et al (2010) Failure to preserve β-cell function with mycophenolate mofetil and daclizumab combined therapy in patients with new-onset type 1 diabetes. Diab Care 33 (4):826–832

Grazia Roncarolo M, Gregori S et al (2006) Interleukin-10-secreting type 1 regulatory T cells in rodents and humans. Immunol Rev 212(1):28–50

Grinberg-Bleyer Y, Baeyens A et al (2010) IL-2 reverses established type 1 diabetes in NOD mice by a local effect on pancreatic regulatory T cells. J Exp Med 207(9):1871–1878

Grossman WJ, Verbsky JW et al (2004a) Human T regulatory cells can use the perforin pathway to cause autologous target cell death. Immunity 21(4):589–601

Grossman WJ, Verbsky JW et al (2004b) Differential expression of granzymes A and B in human cytotoxic lymphocyte subsets and T regulatory cells. Blood 104(9):2840–2848

Haller MJ, Atkinson MA, Schatz D (2005) Type 1 diabetes mellitus: etiology, presentation, and management. Pediatr Clin North Am 52:1553–1578

Han S, Donelan W et al (2011) Autoantigen-specific immunotherapy. In: Wagner D (ed) Type 1 diabetes—pathogenesis, genetics and immunotherapy. ISBN: 978-953-307-362-0. doi:10. 5772/24479. http://www.intechopen.com/books/type-1-diabetes-pathogenesis-genetics-and-immunotherapy/autoantigen-specific-immunotherapy (InTech)

Hao W, Davis C et al (1999) Plasmapheresis and immunosuppression in stiff-man syndrome with type 1 diabetes: a 2-year study. J Neurol 246(8):731–735

Harrison LC (2012) Insulin-specific vaccination for type 1 diabetes: a step closer? Hum Vaccin Immunother 8(6):834–837

Herold KC, Gitelman S et al (2009) Treatment of patients with new onset Type 1 diabetes with a single course of anti-CD3 mAb Teplizumab preserves insulin production for up to 5 years. Clin Immunol 132:166–173

Herold KC, Bluestone JA et al (1992) Prevention of autoimmune diabetes with nonactivating anti-CD3 monoclonal antibody. Diabetes 41(3):385–391

Hovorka R (2008) The future of continuous glucose monitoring: closed loop. Curr Diab Rev 4:269–279

Hovorka R (2005) Continuous glucose monitoring and closed-loop systems. Diabet Med 23(1):1–12

Hilsted J, Madsbad S et al (1995) Intranasal insulin therapy: the clinical realities. Diabetologia 38 (6):680–684

Harrison LC, Honeyman MC et al (2004) Pancreatic beta-cell function and immune responses to insulin after administration of intranasal insulin to humans at risk for type 1 diabetes. Diab Care 27(10):2348–2355

Harrison LC (2008) Vaccination against self to prevent autoimmune disease: the type 1 diabetes model. Immunol Cell Biol 86(2):139–145

Humar A, Kandaswamy R et al (2000) Decreased surgical risks of pancreas transplantation in the modern era. Ann Surg 231(2):269–275

Huurman VA, van der Meide PE et al (2008) Immunological efficacy of heat shock protein 60 peptide DiaPep277 therapy in clinical type I diabetes. Clin Exp Immunol 152(3):488–497

Hinke SA (2011) Inverse vaccination with islet autoantigens to halt progression of autoimmune diabetes. Drug Dev Res 72(8):788–804

Jacob CO, Aiso S et al (1990) Prevention of diabetes in nonobese diabetic mice by tumor necrosis factor (TNF): similarities between TNF-alpha and interleukin 1. Proc Natl Acad Sci USA 87 (3):968–972

Jamal A, Lipsett M et al (2005) Morphogenetic plasticity of adult human pancreatic islets of Langerhans. Cell Death Differ 12(7):702–712

Jenner M, Bradish G et al (1992) Cyclosporin A treatment of young children with newly-diagnosed Type 1 (insulin-dependent) diabetes mellitus. Diabetologia 35(9):884–888

Kaizer EC, Glaser CL et al (2007) Gene expression in peripheral blood mononuclear cells from children with diabetes. J Clin Endocrinol Metab 92(9):3705–3711

Karlsen AE, Hagopian WA et al (1991) Cloning and primary structure of a human islet isoform of glutamic acid decarboxylase from chromosome 10. Proc Natl Acad Sci USA 88(19):8337–8341

Kaufman D, Erlander M et al (1992) Autoimmunity to two forms of glutamate decarboxylase in insulin-dependent diabetes mellitus. J Clin Invest 89(1):283–292

Kaufman DL, Clare-Salzler M et al (1993) Spontaneous loss of T-cell tolerance to glutamic acid decarboxylase in murine insulin-dependent diabetes. Nature 366:69–72

Kelly WD, Lillehei RC et al (1968) Allotransplantation of the pancreas and duodenum along with the kidney in diabetic nephropathy. Transplantation 6(1):827–837

Kent SC, Chen Y et al (2005) Expanded T cells from pancreatic lymph nodes of type 1 diabetic subjects recognize an insulin epitope. Nature 435(7039):224–228

Keymeulen B, Vandemeulebroucke E et al (2005) Insulin needs after CD3-antibody therapy in new-onset type 1 diabetes. N Engl J Med 352(25):2598–2608

Keymeulen B, Candon S et al (2010a) Transient Epstein-Barr virus reactivation in CD3 monoclonal antibody-treated patients. Blood 115(6):1145–1155

Keymeulen B, Walter M et al (2010b) Four-year metabolic outcome of a randomised controlled CD3-antibody trial in recent-onset type 1 diabetic patients depends on their age and baseline residual beta cell mass. Diabetologia 53(4):614–623

Kroon E, Martinson LA et al (2008) Pancreatic endoderm derived from human embryonic stem cells generates glucose-responsive insulin-secreting cells in vivo. Nat Biotechnol 26(4):443–452

Klinke DJ (2008) Extent of beta cell destruction is important but insufficient to predict the onset of type 1 diabetes mellitus. PLoS ONE 3(1):e1374

Kodama S, Davis M et al (2005) The therapeutic potential of tumor necrosis factor for autoimmune disease: a mechanistically based hypothesis. Cell Mol Life Sci 62(16):1850–1862

Koulmanda M, Bhasin M et al (2012) The Role of TNF-α in Mice with Type 1-and 2-Diabetes. PLoS ONE 7(5):e33254

Krishna M, Huissoon A (2011) Clinical immunology review series: an approach to desensitization. Clin Exp Immunol 163(2):131–146

Kabelitz D, Geissler EK et al (2008) Toward cell-based therapy of type I diabetes. Trends Immunol 29(2):68–74

Keselowsky BG, Xia CQ et al (2011) Multifunctional dendritic cell-targeting polymeric microparticles- Engineering new vaccines for type 1 diabetes. Hum Vaccines 7(1):37–44

Kim SJ, Nian C et al (2009) Dipeptidyl peptidase IV inhibition with MK0431 improves islet graft survival in diabetic NOD mice partially via T-cell modulation. Diabetes 58(3):641–651

Kamal M, Al Abbasy AJ et al (2006) Effect of nicotinamide on newly diagnosed type 1 diabetic children. Acta Pharmacol Sin 27(6):724–727

Lahl K, Loddenkemper C et al (2007) Selective depletion of Foxp3 + regulatory T cells induces a scurfy-like disease. J Exp Med 204(1):57–63

Larsen CM, Faulenbach M et al (2007) Interleukin-1–receptor antagonist in type 2 diabetes mellitus. N Engl J Med 356(15):1517–1526

Lazar L, Ofan R et al (2007) Heat-shock protein peptide DiaPep277 treatment in children with newly diagnosed type 1 diabetes: a randomised, double-blind phase II study. Diab Metab Res Rev 23(4):286–291

Lenschow DJ, Zeng Y et al (1992) Long-term survival of xenogenic pancreatic islet grafts induced by CTLA4Ig. Science 257:789–792

Lenschow DJ, Herold KC et al (1996) CD28/B7 regulation of Th1 and Th2 subsets in the development of autoimmune diabetes. Immunity 5(3):285–293

Levy LM, Dalakas MC et al (1999) The stiff-person syndrome: an autoimmune disorder affecting neurotransmission of γ-aminobutyric acid. Ann Intern Med 131(7):522–530

Li A, Escher A (2011) Immunotherapy for type 1 diabetes-preclinical and clinical trials. In: Wagner D (ed) Type 1 diabetes—pathogenesis, genetics and immunotherapy. ISBN: 978-953-307-362-0. doi:10.5772/22049. http://www.intechopen.com/books/type-1-diabetes-pathogenesis-genetics-and-immunotherapy/immunotherapy-for-type-1-diabetes-preclinical-and-clinical-trials(InTech)

Limbert C, Päth G et al (2008) Beta-cell replacement and regeneration: strategies of cell-based therapy for type 1 diabetes mellitus. Diab Res Clin Pract 79(3):389–399

Lipsett M, Aikin R et al (2006) Islet neogenesis: a potential therapeutic tool in type 1 diabetes. Int J Biochem Cell Biol 38(5–6):715–720

Liu W, Putnam AL et al (2006) CD127 expression inversely correlates with FoxP3 and suppressive function of human CD4 + T reg cells. J Exp Med 203(7):1701–1711

Lohmann T, Hawa M et al (2000) Immune reactivity to glutamic acid decarboxylase 65 in stiff-man syndrome and type 1 diabetes mellitus. Lancet 356(9223):31–35

Long SA, Rieck M et al (2012) Rapamycin/IL-2 combination therapy in patients with type 1 diabetes augments tregs yet transiently impairs β-cell function. Diabetes 61(9):2340–2348

Looney R (2005) B cells as a therapeutic target in autoimmune diseases other than rheumatoid arthritis. Rheumatology (Oxford) 44(Suppl 2):ii(13)–(17)

Ludvigsson J, Heding L et al (1983) Plasmapheresis in the initial treatment of insulin-dependent diabetes mellitus in children. Br Med J (Clin Res Ed) 286(6360):176–178

Ludvigsson J, Faresjö M et al (2008) GAD treatment and insulin secretion in recent-onset type 1 diabetes. N Engl J Med 359(18):1909–1920

Ludvigsson J (2009) The role of immunomodulation therapy in autoimmune diabetes. J Diab Sci Technol 3(2):320–330

Ludvigsson J, Hjorth M et al (2011) Extended evaluation of the safety and efficacy of GAD treatment of children and adolescents with recent-onset type 1 diabetes: a randomised controlled trial. Diabetologia 54(3):634–640

Ludvigsson J, Krisky D et al (2012) GAD65 Antigen therapy in recently diagnosed type 1 diabetes mellitus. N Engl J Med 366(5):433–442

Machen J, Harnaha J et al (2004) Antisense oligonucleotides down-regulating costimulation confer diabetes-preventive properties to nonobese diabetic mouse dendritic cells. J Immunol 173 (7):4331–4341

Marek-Trzonkowska N, Mysliwiec M et al (2012) Administration of CD4 + CD25highCD127-regulatory T cells preserves beta-cell function in type 1 diabetes in children. Diab Care 35 (9):1817–1820

Martin F, Chan AC (2006) B cell immunobiology in disease: evolving concepts from the clinic. Annu Rev Immunol 24:467–496

Masteller EL, Warner MR et al (2005) Expansion of functional endogenous antigen-specific CD4 + CD25 + regulatory T cells from nonobese diabetic mice. J Immunol 175(5):3053–3059

Mastrandrea L, Yu J et al (2009) Etanercept treatment in children with new-onset type 1 diabetes: pilot randomized, placebo-controlled, double-blind study. Diab Care 32(7):1244–1249

Matthews J, Staeva T et al (2010) Developing combination immunotherapies for type 1 diabetes: recommendations from the ITN–JDRF type 1 diabetes combination therapy assessment group. Clin Exp Immunol 160(2):176–184

Mercer F, Unutmaz D (2009) The biology of FoxP3: a key player in immune suppression during infections, autoimmune diseases and cancer. Adv Exp Med Biol 665:47–59

Millington OR, Mowat AMI et al (2004) Induction of bystander suppression by feeding antigen occurs despite normal clonal expansion of the bystander T cell population. J Immunol 173 (10):6059–6064

Miyara M, Sakaguchi S (2007) Natural regulatory T cells: mechanisms of suppression. Trends Mol Med 13(3):108–116

Morelli AE, Thomson AW (2007) Tolerogenic dendritic cells and the quest for transplant tolerance. Nat Rev Immunol 7(8):610–621

Mukherjee R, Chaturvedi P et al (2003) CD4 + CD25 + regulatory T cells generated in response to insulin B:9-23 peptide prevent adoptive transfer of diabetes by diabetogenic T cells. J Autoimmun 21(3):221–237

Nakayama M, Abiru N et al (2005) Prime role for an insulin epitope in the development of type 1 diabetes in NOD mice. Nature 435(7039):220–223

Nanji SA, Shapiro A (2006) Advances in pancreatic islet transplantation in humans. Diab Obes Metab 8(1):15–25

Näntö-Salonen K, Kupila A et al (2008) Nasal insulin to prevent type 1 diabetes in children with HLA genotypes and autoantibodies conferring increased risk of disease: a double -blind, randomised controlled trial. Lancet 372(9651):1746–1755

Narayan VKM, Zhang P et al (2006) Diabetes: the pandemic and potential solutions. In: Jamson DT, Breman JG et al (eds) Disease control priorities in development countries. Oxford University Press, World Bank, Washington, pp 591–603

Orban T, Farkas K et al (2010) Autoantigen-specific regulatory T cells induced in patients with type 1 diabetes mellitus by insulin B-chain immunotherapy. J Autoimm 34(4):408–415

Orban T, Bundy B et al (2011) Co-stimulation modulation with abatacept in patients with recent-onset type 1 diabetes: a randomised, double-blind, placebo-controlled trial. Lancet 378 (9789):412–419

Parker MJ, Xue S et al (2009) Immune depletion with cellular mobilization imparts immunoregulation and reverses autoimmune diabetes in nonobese diabetic mice. Diabetes 58(10):2277–2284

Peakman M, Von Herrath M (2010) Antigen-specific immunotherapy for type 1 diabetes: maximizing the potential. Diabetes 59(9):2087–2093

Perrin PJ, Scott D et al (1995) Role of B7: CD28/CTLA-4 in the induction of chronic relapsing experimental allergic encephalomyelitis. J Immunol 154(3):1481–1490

Pescovitz MD, Greenbaum CJ et al (2009) Rituximab, B-lymphocyte depletion, and preservation of beta-cell function. N Engl J Med 361(22):2143–2152

Pescovitz MD, Torgerson TR et al (2011) Effect of rituximab on human in vivo antibody immune responses. J Allergy Clin Immunol 128(6):1295–1302(e1295)

Pickup JC, Zhi ZL et al (2008) Nanomedicine and its potential in diabetes research and practice. Diab Metab Res Rev 24(8):604–610

Powers AC (2008) Insulin therapy versus cell-based therapy for type 1 diabetes mellitus: what lies ahead? Nat Clin Pract Endocrinol Metab 4(12):664–665

Pozzilli P, Pitocco D et al (2000) No effect of oral insulin on residual beta-cell function in recent-onset type I diabetes (the IMDIAB VII). IMDIAB Group. Diabetologia 43(8):1000–1004

Qin S, Cobbold SP et al (1993) "Infectious" transplantation tolerance. Science 259(5097):974–977

Quintana FJ, Carmi P et al (2003) DNA fragments of the human 60-kDa heat shock protein (HSP60) vaccinate against adjuvant arthritis: identification of a regulatory HSP60 peptide. J Immunol 171(7):3533–3541

Rabinovitch A, Suarez-Pinzon WL (2007) Roles of cytokines in the pathogenesis and therapy of type 1 diabetes. Cell Biochem Biophys 48(2–3):159–163

Rabinovitch A, Suarez-Pinzon WL et al (2002) Combination therapy with sirolimus and interleukin-2 prevents spontaneous and recurrent autoimmune diabetes in NOD mice. Diabetes 51(3):638–645

Ragno S, Colston MJ et al (1997) Protection of rats from adjuvant arthritis by immunization with naked DNA encoding for mycobacterial heat shock protein 65. Arthritis Rheum 40(2):277–283

Raz I, Ziegler AG et al (2014) Treatment of recent-onset type 1 diabetic patients with DiaPep277: results of a double-blind, placebo-controlled, randomized phase 3 trial. Diab Care 37(5):1392–1400

Raz I, Avron A et al (2007) Treatment of new-onset type 1 diabetes with peptide DiaPep277 is safe and associated with preserved beta-cell function: extension of a randomized, double-blind, phase II trial. Diab Metab Res Rev 23(4):292–298

Raz I, Elias D et al (2001) [beta]-cell function in new-onset type 1 diabetes and immunomodulation with a heat-shock protein peptide (DiaPep277): a randomised, double-blind, phase II trial. Lancet 358(9295):1749–1753

Renard E (2002) Implantable closed-loop glucose-sensing and insulin delivery: the future for insulin-pump therapy. Curr Opin Pharmacol 2:708–716

Renard E, Schaepelynck BP (2007) Impalntable insulin pumps. A position statement about their clincal use. Diab Metab 33(2):158–166

Renard E (2015) Continous intraperitoneal insulin infusion from implantable pumps. In: Bruttomesso D, Grassi G (eds) Technological advances in the treatment of type 1 diabetes, vol 24., Front DiabetesBasel, Karger, pp 190–209

Rewers M, Gottlieb P (2009) Immunotherapy for the prevention and treatment of type 1 diabetes. Diab Care 32(10):1769–1782

Rigby MR, Trexler AM et al (2008) CD28/CD154 blockade prevents autoimmune diabetes by inducing nondeletional tolerance after effector t-cell inhibition and regulatory T-cell expansion. Diabetes 57(10):2672–2683

Roep B, Atkinson M (2004) Animal models have little to teach us about type 1 diabetes: 1. In support of this proposal. Diabetologia 47(10):1650–1656

Rosenberg L, Lipsett M et al (2004) A pentadecapeptide fragment of islet neogenesis-associated protein increases beta-cell mass and reverses diabetes in C57BL/6 J mice. Ann Surg 240(5):875–884

Rother KI, Spain LM et al (2009) Effects of exenatide alone and in combination with daclizumab on β-cell function in long-standing type 1 diabetes. Diab Care 32(12):2251–2257

Ryan EA, Lakey JRT et al (2001) Clinical outcomes and insulin secretion after islet transplantation with the Edmonton protocol. Diabetes 50(4):710–719

Ryan EA, Paty BW et al (2005) Five-year follow-up after clinical islet transplantation. Diabetes 54(7):2060–2069

Sai P, Damgé C et al (1996) Prophylactic oral administration of metabolically active insulin entrapped in isobutylcyanocrylate nanocapsules reduces the incidence of diabetes in nonobese diabetic mice. J Autoimmun 9(6):713–722

Salomon B, Lenschow DJ et al (2000) B7/CD28 costimulation is essential for the homeostasis of the CD4 + CD25 + immunoregulatory T cells that control autoimmune diabetes. Immunity 12(4):431–440

Salomon B, Bluestone JA (2001) Complexities of CD28/B7: CTLA-4 costimulatory pathways in autoimmunity and transplantation. Annu Rev Immunol 19(1):225–252

Sakaguchi S (2004) Naturally arising CD4 + regulatory T cells for immunologic self-tolerance and negative control of immune responses. Annu Rev Immunol 22:531–562

Saudek F, Havrdova T et al (2004) Polyclonal anti-T-cell therapy for type 1 diabetes mellitus of recent onset. Rev Diab Stud 1(2):80–88

Schloot NC, Meierhoff G et al (2007) Effect of heat shock protein peptide DiaPep277 on beta-cell function in paediatric and adult patients with recent-onset diabetes mellitus type 1: two prospective, randomized, double-blind phase II trials. Diab Metab Res Rev 23(4):276–285

Schreiber SL, Crabtree GR (1992) The mechanism of action of cyclosporin A and FK506. Immunol Today 13(4):136–14

Schneider DA, Sarikonda G et al (2012) Combination therapy with InsB9-23 peptide immunization and CTLA4-IgG does not reverse diabetes in NOD mice. Clin Immunol 142 (3):402–403

Selmani Z, Naji A et al (2008) Human leukocyte antigen G5 secretion by human mesenchymal stem cells is required to suppress T lymphocyte and natural killer function and to induce CD4 + CD25highFOXP3 + regulatory T cells. Stem Cells 26(1):212–222

Serreze DV, Silveira PA (2003) The role of B lymphocytes as key antigen-presenting cells in the development of T cell-mediated autoimmune type 1 diabetes. Curr Dir Autoimmun 6:212–227

Sesardic D, Rijpkema S et al (2007) New adjuvants: EU regulatory developments. Expert Rev Vaccines 6(5):849–861

Sherry N, Hagopian W et al (2011) Teplizumab for treatment of type 1 diabetes (Protégé study): 1-year results from a randomised, placebo-controlled trial. Lancet 378(9790):487–497

Simon G, Parker M et al (2008) Murine antithymocyte globulin therapy alters disease progression in NOD mice by a time-dependent induction of immunoregulation. Diabetes 57(2):405–414

Shapiro AM, Ricordi C et al (2006) International trial of the Edmonton protocol for islet transplantation. N Engl J Med 355(13):1318–1330

Sherry NA, Chen W et al (2007) Exendin-4 improves reversal of diabetes in NOD mice treated with anti-CD3 monoclonal antibody by enhancing recovery of beta-cells. Endocrinol 148(11): 5136–5144

Skyler J, Krischer J et al (2005) Effects of oral insulin in relatives of patients with type 1 diabetes: the diabetes prevention trial-type 1. Diab Care 28(5):1068–1076

Skyler JS (2008) Update on worldwide efforts to prevent type 1 diabetes. Ann NY Acad Sci 1150 (1):190–196

Slobodan C, Christian B et al (2011) Antigen-based immune therapeutics for type 1 diabetes: magic bullets or ordinary blanks? Clin Dev Immunol 2011:286248

SoRelle JA, Naziruddin B (2011) Beta cell replacement therapy. In: Wagner D (ed) Type 1 diabetes—pathogenesis, genetics and immunotherapy. ISBN: 978-953-307-362-0. doi:10. 5772/22283. http://www.intechopen.com/books/type-1-diabetes-pathogenesis-genetics-and-immunotherapy/beta-cell-replacement-therapy (InTech)

Soria B, Roche E et al (2000) Insulin-secreting cells derived from embryonic stem cells normalize glycemia in streptozotocin-induced diabetic mice. Diabetes 49(2):157–162

Spoerl S, Li XC (2011) Regulatory T cells and the quest for transplant tolerance. Discov Med 11 (56):25–34

Sreenan S, Pick AJ et al (1999) Increased beta-cell proliferation and reduced mass before diabetes onset in the nonobese diabetic mouse. Diabetes 48(5):989–996

Steinman RM, Hawiger D et al (2003) Tolerogenic dendritic cells. Annu Rev Immunol 21(1): 685–711

Steinman RM (2008) Dendritic cells in vivo: a key target for a new vaccine science. Immunity 29:319–324

Stiller C, Dupre J et al (1984) Effects of cyclosporine immunosuppression in insulin-dependent diabetes mellitus of recent onset. Science 223(4643):1362–1367

Suarez-Pinzon WL, Lakey JRT et al (2008) Combination therapy with glucagon-like peptide-1 and gastrin induces-cell neogenesis from pancreatic duct cells in human islets transplanted in immunodeficient diabetic mice. Cell Transplant 17(6):631–640

Sumpter KM, Adhikari S et al (2011) Preliminary studies related to anti-interleukin-1beta therapy in children with newly diagnosed type 1 diabetes. Pediatr Diab 12(7):656–667

Sun Y, Chen L et al (2007) Differentiation of bone marrow-derived mesenchymal stem cells from diabetic patients into insulin-producing cells in vitro. Chin Med J 120(9):771–776

Suntharalingam G, Perry MR et al (2006) Cytokine storm in a phase 1 trial of the anti-CD28 monoclonal antibody TGN1412. N Engl J Med 355(10):1018–1028

Sutherland DER, Gruessner A et al (2004) Beta-cell replacement therapy (pancreas and islet transplantation) for treatment of diabetes mellitus: an integrated approach. Tranplant Proc 36 (6):1697–1699

Tang Q, Henriksen KJ et al (2003) Cutting edge: CD28 controls peripheral homeostasis of CD4 + CD25 + regulatory T cells. J Immunol 171(7):3348–3352

Tang Q, Henriksen KJ et al (2004) In vitro–expanded antigen-specific regulatory T cells suppress autoimmune diabetes. J Exp Med 199(11):1455–1465

Tarbell KV, Petit L et al (2007) Dendritic cell–expanded, islet-specific CD4 + CD25 + CD62L + regulatory T cells restore normoglycemia in diabetic NOD mice. J Exp Med 204(1):191–201

Thomas D, Zaccone P et al (2005) The role of regulatory T cell defects in type 1 diabetes and the potential of these cells for therapy. Rev Diab Stud 2(1):9–18

Thomas HE, Irawaty W et al (2004) IL-1 receptor deficiency slows progression to diabetes in the NOD mouse. Diabetes 53:113–121

Thomson JA, Itskovitz-Eldor J et al (1998) Embryonic stem cell lines derived from human blastocysts. Science 282(5391):1145–1147

Thornton AM, Piccirillo CA et al (2004) Activation requirements for the induction of CD4 + CD25 + T cell suppressor function. Eur J Immunol 34(2):366–376

Thrower S, James L et al (2009) Proinsulin peptide immunotherapy in type 1 diabetes: report of a first-in-man Phase I safety study. Clin Exp Immunol 155(2):156–165

Tian J, Atkinson MA et al (1996) Nasal administration of glutamate decarboxylase (GAD65) peptides induces Th2 responses and prevents murine insulin-dependent diabetes. J Exp Med 183(4):1561–1567

Tisch R, Wang B et al (1999) Induction of glutamic acid decarboxylase 65-specific Th2 cells and suppression of autoimmune diabetes at late stages of disease is epitope dependent. J Immunol 163(3):1178–1187

Trzonkowski P, Bieniaszewska M et al (2009) First-in-man clinical results of the treatment of patients with graft versus host disease with human ex vivo expanded CD4 + CD25 + CD127 − T regulatory cells. Clin Immunol 133(1):22–26

Tuccinardi D, Fioriti E et al (2011) DiaPep277 peptide therapy in the context of other immune intervention trials in type 1 diabetes. Expert Opin Biol Ther 11(9):1233–1240

Ugrasbul F, Moore WV et al (2008) Prevention of diabetes: effect of mycophenolate mofetil and anti-CD25 on onset of diabetes in the DRBB rat. Pediatr Diab 9(6):596–601

Uibo R, Lernmark A (2008) GAD65 autoimmunity-clinical studies. Adv Immunol 100:39–78

Vignali DA, Collison LW et al (2008) How regulatory T cells work. Nat Rev Immunol 8(7): 523–532

Vija L, Farge D et al (2009) Mesenchymal stem cells: stem cell therapy perspectives for type 1 diabetes. Diab Metab 35(2):85–93

Voltarelli JC, Couri CE et al (2007) Autologous nonmyeloablative hematopoietic stem cell transplantation in newly diagnosed type 1 diabetes mellitus. JAMA 297(14):1568–1579

Waldron-Lynch F, Herold KC (2011) Immunomodulatory therapy to preserve pancreatic β-cell function in type 1 diabetes. Nat Rev Drug Discov 10(6):439–452

Walter M, Philotheou A et al (2009) No effect of the altered peptide ligand NBI-6024 on beta-cell residual function and insulin needs in new-onset type 1 diabetes. Diab Care 32(11):2036–2040

Werdelin O, Cordes U et al (1998) Aberrant expression of tissue-specific proteins in the thymus: a hypothesis for the development of central tolerance. Scand J Immunol 47(2):95–100

Wherrett DK, Bundy B et al (2011) Antigen-based therapy with glutamic acid decarboxylase (GAD) vaccine in patients with recent-onset type 1 diabetes: a randomised double-blind trial. Lancet 378(9788):319–327

Yang XD, Tisch R et al (1994) Effect of tumor necrosis factor alpha on insulin-dependent diabetes mellitus in NOD mice. I. The early development of autoimmunity and the diabetogenic process. J Exp Med 180(3):995–1004

Yi H, Zhen Y et al (2006) The phenotypic characterization of naturally occurring regulatory CD4 + CD25 + T cells. Cell Mol Immunol 3(3):189–195

Zhang ZJ, Davidson L et al (1991) Suppression of diabetes in nonobese diabetic mice by oral administration of porcine insulin. Proc Natl Acad Sci USA 88(22):10252–10256

Zhao Z, Leong KW (1996) Controlled delivery of antigens and adjuvants in vaccine development. J Pharm Sci 85:1261–1270

Zheng XX, Steele AW et al (1999) IL-2 receptor-targeted cytolytic IL-2/Fc fusion protein treatment blocks diabetogenic autoimmunity in nonobese diabetic mice. J Immunol 163 (7):4041–4048

Zheng Y, Rudensky AY (2007) Foxp3 in control of the regulatory T cell lineage. Nat Immunol 8 (5):457–462

Zimmermann H, Zimmermann D et al (2005) Towards a medically approved technology for alginate-based microcapsules allowing long-term immunoisolated transplantation. J Mater Sci Mater Med 16(6):491–501

Zulewski H (2006) Stem cells with potential to generate insulin-producing cells in man. Swiss Med Wkly 136(Suppl 155):60S–67S

Printed in the United States
By Bookmasters